Willis's
Elements of
Quantity Surveyin

Also of interest

Specification Writing for Architects and Surveyors
Eleventh Edition
Christopher J. Willis and J. Andrew Willis
ISBN: 978-0-632-04206-7

Willis's Practice and Procedure for the Quantity Surveyor
Twelfth Edition
Allan Ashworth and Keith Hogg
ISBN: 978-1-4051-4578-7

Willis's
Elements of
Quantity Surveying

Tenth Edition

Sandra Lee
MSc, FRICS, MCIOB

William Trench
FRICS

Andrew Willis
BSc, FRICS, DipArb, FCIArb

Blackwell
Publishing

Blackwell Publishing Editorial offices:
Blackwell Publishing Ltd, 9600 Garsington Road, Oxford OX4 2DQ, UK
  Tel: +44 (0)1865 776868
Blackwell Publishing Inc., 350 Main Street, Malden, MA 02148-5020, USA
  Tel: +1 781 388 8250
Blackwell Publishing Asia Pty Ltd, 550 Swanston Street, Carlton, Victoria 3053, Australia
  Tel: +61 (0)3 8359 1011

Ninth edition published by Blackwell Science 1998
Tenth edition published 2005 by Blackwell Publishing Ltd
4   2008

ISBN: 978-1-4051-2563-5

Library of Congress Cataloging-in-Publication Data
Lee, Sandra, MSc, FRICS, MCIOB
    Willis's elements of quantity surveying / Sandra Lee, Andrew Willis, William Trench. – 10th ed.
        p. cm.
    Rev. ed. of: Willis's elements of quantity surveying / Andrew Willis, William Trench. 1998.
    Includes index.
    ISBN: 978-1-4051-2563-5
    1. Building – Estimates.   I. Willis, Andrew.   II. Trench, William, F.R.I.C.S.   III. Willis, Andrew.
  Willis's elements of quantity surveying.   IV. Title.

    TH435.W685 2005
    692'.5–dc22

                                                                            2004062751

A catalogue record for this title is available from the British Library

Set in 10.5/13pt Times
by DP Photosetting, Aylesbury, Bucks
Printed and bound in India by Replika Press Pvt. Ltd

The publisher's policy is to use permanent paper from mills that operate a sustainable forestry policy, and which has been manufactured from pulp processed using acid-free and elementary chlorine-free practices. Furthermore, the publisher ensures that the text paper and cover board used have met acceptable environmental accreditation standards.

For further information on Blackwell Publishing, visit our website:
www.blackwellpublishing.com

To the memory of Arthur and Christopher Willis

Contents

Preface

In recent years the use of the bill of quantities has somewhat declined and today it is only one of a variety of options open to the industry for the procurement of construction contracts. Nevertheless, the skills of measurement are still very much required in some form or other under most procurement routes. The basic structure of the book generally follows that of previous editions, setting down the measurement process from first principles and assuming the reader is coming fresh to the subject. The text has been amended to reflect modern terminology. The examples have been updated to reflect today's construction technology and an additional example of a simple steel frame structure has been included.

Whilst it is recognised that modern computerised measurement techniques utilising standard descriptions might appear far removed from traditional taking-off, it is only by fully grasping such basic principles of measurement that they can be adapted and applied to alternative systems. It is for this reason that the examples continue to be written in traditional form.

A slightly different approach to measurement is required when using a computer. Guidance on adapting the traditional principles to a range of software is now included.

As previously, the book opens with some practical guidance on measurement generally and the application of mensuration. Specific examples of measurement for elements of a simple building follow and the book ends with an outline of the bill preparation process.

Sandra Lee
William Trench

This book was first published in 1935 and my grandfather Arthur J. Willis, the original author, said in the preface to that first edition that it was intended 'to be a book giving everything in its simplest form and to assist a student to a good grounding in first principles'. Whilst each successive edition has been brought up to date, we have always striven to maintain the guiding principles he set. In 1968 my father joined my grandfather in joint authorship and on his death in 1983 he took on the work, assisted by Don Newman. William Trench then worked with me on the ninth edition. The guiding principles laid down 70 years ago remain as relevant today as they were then, and have been maintained in this latest edition. Whilst the role of the quantity surveyor is subject to continual change, I am sure that students will find this book as useful as their predecessors have.

Andrew Willis

Acknowledgement

We are indebted to Ruth Pearson, who prepared the drawings. Her efforts are gratefully acknowledged.

Table of Examples

Abbreviations

a.b.	as before
a.b.d.	as before described
agg.	aggregate
BS	British Standard
CAWS	Common Arrangement of Work Sections
c/c	centres
ddt	deduct
diam.	diameter
d.p.c.	damp proof course
d.p.m.	damp proof membrane
EDI	Electronic Data Interchange
e.w.s.	earth work support
ex	out of
hw.	hardwood
JCT	Joint Contracts Tribunal
n.t.s.	not to scale
PC	Prime Cost
MC	Measurement Code
n.e.	not exceeding
r.c.	reinforced concrete
r.w.p.	rainwater pipe
SMM	Standard Method of Measurement
sw.	softwood
swg	standard wire gauge

Chapter 1
Introduction

The modern quantity surveyor

The training and knowledge of the quantity surveyor have enabled the role of the profession to evolve over time into new areas, and the services provided by the modern quantity surveyor now cover all aspects of procurement, contractual and project cost management. This holds true whether the quantity surveyor works as a consultant or whether employed by a contractor or sub-contractor. Whilst the importance of this expanded role cannot be emphasised enough, success in carrying it out stems from the traditional ability of the quantity surveyor to measure and value. It is on the aspect of measurement that this book concentrates.

Method of analysing cost

It is evident that if a building is divided up into its constituent parts, and the cost of each part can be estimated, an estimate can be compiled for the whole work. It was found in practice that by making a 'schedule' setting out the quantity of each item of work for a project, the labour and material requirements for these could be more readily assessed. This schedule is often a bill of quantities which when priced can provide a total sum for a project. It must not be forgotten that a traditional bill of quantities only produces an estimate. It is prepared and priced before the erection of the building and gives the contractor's estimated cost. Such an estimated cost, however, under the most commonly used construction contracts, becomes a tender and a definite price for which the contractor agrees to carry out the work as set out in the bill. The bill must, therefore, completely represent the proposed work so that a serious discrepancy between actual and estimated cost does not arise.

Origin of the bill of quantities

Competitive tendering is one of the basic principles of most classes of business, and if competitors are given comprehensive details of the requirements it should be fair to all concerned. However, when tendering, builders found that considerable work was involved in making detailed calculations and measurements to form the basis for a tender. They realised that by getting together and employing one person to make these calculations and measurements for them all, a considerable saving would be made in their overhead charges. They began to arrange for this to be done, each including the surveyor's fee for preparing the bill of quantities in their tender, and the successful competitor paying. Each competing builder was provided with the same bill of quantities which could then be priced in a comparatively short space of time. It was not long before the position was realised by architect and employer. Here the employer was paying indirectly for the quantity surveyor through the builder, whereas the surveyor could be used as a consultant if a direct appointment was made. In this way the quantity surveyor began to get the authority of the employer and was used to prepare a bill of quantities for tendering purposes.

The measurement process

The main purpose of a bill of quantities is therefore for tendering. Each contractor tendering for a project is able to price the work on exactly the same information with a minimum of effort. This gives rise to the fairest type of competition.

Despite the demise of the bills of quantities on large projects, over 50% of the value of all building work in the UK is still let using lump-sum contracts with firm or approximate quantities. Most other procurement routes, such as design and build and management contracting in its various forms, also involve quantification of the work in some form or other, either by the main contractor, subcontractor or package contractor, and therefore the measurement process continues to be of importance.

Computerised and other alternative measurement systems have become more and more widely used. However, it is only by having a detailed understanding of the traditional method of setting down dimensions and framing descriptions that such systems can be fully understood and properly utilised.

Attributes of a quantity surveyor

What then are the desired skills of a good quantity surveyor? An ability to describe clearly, fully and precisely the requirements of the designers and arrange the bill of quantities so that the contractor's estimator can quickly, easily and accurately arrive at the estimated cost of the work is essential. This being so, it is obviously important that the surveyor should be able to write clearly in language that will not be misunderstood, and must have a sound knowledge of building materials and construction and of customs prevailing in the industry. Moreover, the surveyor must be careful and accurate in making calculations, have a systematic and orderly mind and be able to visualise the drawings and details.

Divisions of bill preparation

The traditional preparation of a bill of quantities divides itself into two distinct stages:

(1)   The measurement of the dimensions and the compilation of the descriptions from the drawings and specification. This process is commonly known as *taking-off*.
(2)   The preparation of the bill. This involves the calculation of volumes, areas, etc. (*squaring the dimensions*). Traditionally, this was followed by entering the descriptions and the squared dimensions on an abstract (*abstracting*) to collect similar items together and present them in a recognised bill order. From this abstract the draft bill was written (*billing*).

With the virtual disappearance of the worker-up, who was traditionally responsible for all calculations, abstracting and billing, and the introduction of the calculator for squaring and checking dimensions, the abstracting stage gave way to a system of slip sorting known as *cut and shuffle*, whereby the original dimensions and descriptions were copied and then cut into slips and sorted into bill order.

  More recently, through the utilisation of computerised systems, the various stages have become more integrated. The facility now exists for direct input of quantities and formulation of descriptions through the use of standard libraries of descriptions, and the lengthy collating and bill preparation processes is carried out automatically. It should be noted that there is often still the need to produce preliminaries and preambles separately and input uncommon items (*rogue* items) that are peculiar to the particular project. Checking total quantities and careful editing of the bill are still required to identify any data entry problems.

Quantities as part of the contract

Where a contract with quantities is used for a project the bill of quantities forms one of the contract documents, the contract providing that the quantity of work comprised in the contract shall be that set out in the bill of quantities. In such a case the contractor is expected to carry out and the employer to pay for neither more nor less than the quantities given, an arrangement that is fair to both parties. Thus it will be seen how important accuracy is in the preparation of the bill, and how a substantial error might lead an employer to enter into a contract that involved a sum considerably beyond that contemplated.

So that contractors appreciate how the quantities have been prepared and what is included in each item, the quantity surveyor uses the rules from a standard method of measurement. In the UK, the current standard method used for measuring building works is SMM7.

Standard Method of Measurement (SMM)

The *Standard Method of Measurement*, 7th Edition, (SMM7) and the *Code of Measurement Practice (MC)*, both agreed by the Royal Institution of Chartered Surveyors and the former Building Employers Confederation (now the Construction Confederation) set out rules for the measurement and description of building work. SMM7 is a document that provides not only a uniform basis for measuring building work but also embodies the essentials of good practice. If all bills of quantities are prepared in accordance with these rules then all parties concerned are aware of what is included and what is to be assumed. Contractors therefore tender on an equal basis and their tenders can be more readily compared and evaluated. Without the use of such a set of rules the quality of bills of quantities can vary widely. The *Code of Measurement Practice* is a companion volume which clarifies and explains the rules contained in the SMM.

Other standard methods of measurement have been produced for such specific activities as the measurement of civil engineering work and industrial engineering work; however, this book concentrates solely on SMM7.

Contractor produced quantities and estimates

If, however, the bill of quantities is not part of the contract, as for example when a contractor prepares a tender from drawings and specification, the risk of errors in the quantities is taken by the contractor. When the quantities are prepared by an independent surveyor it would be unfair that this risk should fall on the contractor.

The subject of quantity surveying is dealt with in this book chiefly from the point of view of preparing a bill of quantities, but the ability to prepare quantities is also very necessary to the contractor for the compilation of tenders where quantities are not supplied, which is common in design and build and small contracts. A contractor may well produce quantities for an estimate including only the main items of work and not all of the items that would have been measured using SMM7. The descriptions would be shorter and usually the pricing is worked out alongside the quantities, thus avoiding abstracting and billing. The contractor's estimate is solely for internal pricing. If mistakes are made or short cuts are taken which lead to errors, the contractor alone suffers. A bill of quantities will be interpreted by a number of contractors in competition and it must therefore be complete and a suitable basis for a contract. The same general principles of measurement will apply in both cases, but the contractor is free to adapt them to the company's needs, whereas the quantity surveyor is bound by standard rules in the interests of all contractors tendering. Nevertheless, the contractor's surveyor must be able to check a bill of quantities and measure variations on the basis of that bill. It is therefore essential that the contractor's surveyor should understand how the bill is prepared, and there should be no difficulty in adapting this knowledge to suit the somewhat different requirements when preparing quantities for a contractor's estimate.

Differences of custom

It must be understood that, as a good deal of the subject matter of this book is concerned with method and procedure, suggestions made must not be taken as invariable rules. Surveyors will have in many cases their own customs and methods of working which may differ from those given here, and which may be equally good, or in their view better. The procedure advocated is put forward as being reasonable and based on practice. Every effort has been made to explain reasons for suggestions, so that they can be balanced against any alternative proposed. Furthermore, all rules must be adapted to suit any particular circumstances of the project in hand.

Co-ordinated project information (CPI)

Following pressure from the construction industry to have codes of practice for producing project information that was co-ordinated, easy to search and enabled inadequacies to be identified, three conventions were produced in 1987. These were:

(1) Common arrangement of works sections (CAWS)
(2) Drawings code – code of practice for production drawings
(3) Specifications code – code of practice for producing project specifications.

SMM7 was published in 1988 utilising CAWS and was a co-ordinated document.

Common Arrangement of Work sections (CAWS)

This is a system devised to define an efficient and generally acceptable arrangement for specifications and bills of quantities. The main advantages are:

- Easier distribution of information
- A more effective reading together of documents
- Greater consistency of content and description.

CAWS includes some 300 work sections commonly encountered in the building industry. They vary widely in their scope and nature, reflecting the extensive range of products and materials that now exist for use by contractors, subcontractors and specialists. CAWS is arranged with items set out in three levels. The third and lowest of these levels consists of the work sections themselves; levels one and two are the headings under which they are grouped. The SMM, being a coordinated document, is drafted using these same three levels. An example of the levels is shown below:

Level 1	Group	Groundwork	D
Level 2	Sub-group	Piling	D30
Level 3	Work sections	Driven sheet piling	D30.27

The coding of the levels in this way has also guided the production of libraries of descriptions used with computer software for bill production. This will be covered in slightly more detail later.

Method of study

The student is advised in the first place to study Chapter 2 in order to grasp thoroughly the form in which dimensions are usually written. The student is assumed to have knowledge of elementary building construction and simple mensuration and trigonometry; a student who is weak in these subjects should study them before proceeding further with measurement. Chapter 4 gives some information on, and examples of, the practical application of mensuration most commonly met with in quantity surveying. In Chapter 5 are collected a number of notes on general procedure which should be read before attempting to study actual examples of measurement, but to which the student may also find it useful to refer after having made some study of taking-off. Chapters 6 to 17 represent the clear-cut sections into which the taking-off of a small building might be divided, and these should be taken one at a time. The principal clauses of the SMM and the MC applicable are referred to in each chapter and should be studied concurrently. After the chapter has been read, the examples should be worked through. The student should be able to follow every measurement by reference to the drawing.

The examples of taking-off in this book are small isolated parts of what could be the dimensions of a complete building and are not a connected series. When they have been mastered in their isolation it will be much easier to see how they might be expanded and fitted together to make up the dimensions of a complete building.

Chapters 18 and 19 deal with preliminaries and bill preparation, which is more logically dealt with after taking-off as this is a subsequent process. The student may, however, study these chapters simultaneously with those on taking-off.

Examples

The examples are included to illustrate the methods of measurement of a small unit of a building. They assume that full specification clauses would be set out in preambles to the bill (see Chapter 18), and that clauses would be inserted, such as those required by the SMM for keeping excavation free from water, testing drainage, etc.

The dimensions set down in the dimension column when taking-off are given to the nearest two decimal places of a metre in accordance with the guidance given by the SMM (GR3.2). Side casts, or waste calculations, as they are sometimes called, are used to calculate these dimensions, and are given in millimetres to ensure accuracy.

The examples in the chapters are presented in a traditional dimension format, this being considered the best system for a textbook and what the candidate will usually be faced with in the examination room. Abbreviations have been used for deductions where a description sufficient to recognise an item clearly is all that is required. The abbreviations used in the descriptions are listed on the 'Abbreviations' page xii.

Chapter 2
Setting Down Dimensions

The development of computerised measurement and billing systems, each with its own structure for inputting dimensions and calling up descriptions, has made the more traditional procedures less common nowadays. However, it is only by explaining the basic principles of setting down dimensions and descriptions in traditional form, as detailed here and throughout the following chapters, that they can be fully understood and then applied to whatever measurement process is adopted.

Traditional dimension paper

The dimensions are measured from the drawings by the taker-off, using paper ruled as follows:

1	2	3	4	1	2	3	4

The columns (not of course normally numbered) have been numbered here for identification. Column 1 is called the *timesing column*, and its use will be described later. Column 2 is the *dimension column*, in which the measurements are set down as taken from the drawings. Column 3 is the *squaring column*, in which are set out the calculated volumes, areas, etc., of the measurements in column 2. Column 4 is the *description column*, in which is written the description of the work to which the dimensions apply, and on the extreme right-hand side of which (known as *waste*) preliminary calculations and collections are made. There are two sets of columns in the width of a single A4 sheet. No written work should be carried across the central vertical division. There is usually a narrow binding margin (not shown above) on the left of the sheets.

Each dimension sheet should have the name and/or the number of the project written on or, better, stamped on. In addition the title of the section being measured should be included, followed by a number, starting at 1 for each section. The examples measured in the following chapters have been written using only one half of the sheet, the right-hand side being used for explanation.

Form of dimensions

Before going any further it is necessary to understand the dimensions as set down by the taker-off. All dimensions are in one of five forms:

(1)   Cubic measurements
(2)   Square or superficial measurements
(3)   Linear measurements
(4)   Enumerated items
(5)   Items.

These are expressed in the first three cases by setting down the measurements immediately under each other in the dimension column, each separate item being divided from the next by a line, for example:

3.00 2.00 4.00		indicating a cubic measurement 3.00 m long, 2.00 m wide and 4.00 m deep.
3.00 2.00		indicating a superficial measurement 3.00 m long and 2.00 m wide.
3.00		indicating a linear measurement of 3.00 m.

An item to be enumerated is usually indicated in one of the following ways:

	$\underline{4}$	indicating four in number.
	Nr 4	
4/	$\underline{1}$	

Occasionally the SMM requires the insertion of an *item*; this is a description without a measured quantity, e.g. testing the drainage system. The description may, if applicable, contain dimensions, e.g. temporary screens. This requirement is indicated as follows:

	<u>Item</u>	

There is no need to label dimensions *cube, super, lin*, etc., as, if a rule is made always to draw a line under each measurement, it is obvious from the number of entries in the measurement under which category it comes. It is usual to set down the dimensions in the following order:

(1)   Horizontal length
(2)   Horizontal width or breadth
(3)   Vertical depth or height.

Although the order will not affect the calculations of the cubic or square measurement, it is valuable in tracing measurements later if a consistent order is maintained. As will be explained, an incorrect order in a description may even sometimes mislead an estimator in pricing.

Timesing

It often happens that, when the taker-off has written the dimension, it is found that there are several items having the same measurements; to indicate that the measurement is to be multiplied it will be *timesed* as follows:

3/	3.00		indicating that the cubic measurement is to be multiplied by 3.
	2.00		
	4.00		

| 5/ | 3.00 | | indicating that the superficial measurement is to be multiplied by 5. |
| | 2.00 | | |

The timesing figure is kept in the first column and separated from the dimension by a diagonal stroke. An item timesed can be timesed again, each multiplier multiplying everything to the right of it, as follows:

5/3/	3.00		indicating that the cubic measurement having been multiplied by 3, the result is to be multiplied by 5, i.e. the original measurement is multiplied by 15.
	2.00		
	4.00		

2/5/3/	3.00		indicating that the cubic measurement is to be multiplied by 30.
	2.00		
	4.00		

The timesing is done to a linear or enumerated item in just the same way as shown above.

Dotting on

In repeating a dimension, the taker-off may find that it cannot be multiplied but can be added. For instance, given that three items have been measured as follows:

3/	3.00		
	2.00		
	4.00		

To make the train of thought clearer, what is called *dotting on* can be used as follows:

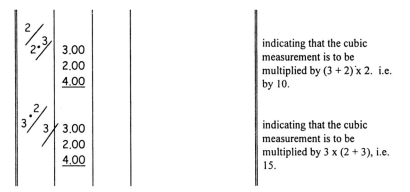

indicating that the cubic measurement is to be multiplied by 3 + 2, i.e. by 5.

The dot is placed below the top figure to avoid any possible confusion with decimals, although these are usually avoided in timesing. Figures dotted on should be lower than the last, just as each one timesed is usually higher; this makes more space available than if they were all written in a horizontal line.

Timesing and dotting on can be combined, as follows:

indicating that the cubic measurement is to be multiplied by (3 + 2) x 2. i.e. by 10.

indicating that the cubic measurement is to be multiplied by 3 x (2 + 3), i.e. 15.

Care must be taken when writing fractions in the timesing column that the line dividing the numerator from the denominator is horizontal, thereby avoiding any confusion with timesing.

Waste calculations

Except in very simple cases, dimensions should not be calculated mentally. Not only will the risk of error be reduced if the calculations are written down, because they will be checked, but another person can readily see the origin of the dimension. These preliminary calculations, known as *waste calculations*, are made on the right-hand side of the description column. They must be written definitely and clearly, and not scribbled as if they were a calculation worked out on scrap paper. The term *waste* used for this part of the column might be thought to imply 'useless' but in fact implies 'a means to an end'. Every effort must be made to commit to writing the train of thought of the taker-off. Waste calculations should be limited to those necessary for the clear setting down of the dimensions by the taker-off, and should not take the place of squaring.

An example of waste calculations is given later in this chapter. These should not be written directly on the drawings. It is, however, necessary for dimensions to be traced from the drawings through any adjustments in the waste calculations to the figures used in the dimension column. They should preferably precede the description. An example of this is included along with the description for excavating a trench at the end of the next section on descriptions.

Alterations in dimensions

Where a dimension has been set down incorrectly and is to be altered, either it should be neatly crossed out and the new dimension written in, or the word *nil* should be written against it in the squaring column to indicate that it is cancelled. Where there are a number of measurements in the dimension column, care must be taken to indicate clearly how far the nil applies; this may be done as follows:

3.00	
4.00	
6.00	
2.00	
7.00	
3.00	NIL
2.60	
5.00	

No attempt should be made to alter figures, e.g. a 2 into a 3, or a 3 into an 8. The figure may appear to have been altered satisfactorily, but may look quite different to another person or when the dimensions are photocopied. Every figure must be absolutely clear, a page a little untidy but with unmistakable figures being far preferable to one where the figures give rise to uncertainties and consequent error. Deletions with correcting fluid should never be made as it is often of value to know what was written in the first instance. It is best to nil entirely and write out again any dimensions that are getting too confused by alterations; but great care is needed in copying dimensions, or mistakes may be made, it being particularly easy to miss copying the timesing. Therefore, where any dimensions are rewritten, they should be checked very carefully against the original.

The descriptions

The description of the item measured is written in the description column, on a level with its associated dimensions as follows (the waste calculation also being shown):

			34 000
			16 000
			2/50 000
			= 100 000
		- 4/215	860
			99 140
			15 050
			12 000
			13 500
			11 700
			4)52 250
			av = 13 060
		Underside	
		of conc. -	12 050
			1 010
	99.14	Excavating trenches	
	1.00	width exceeding	
	1.01	0.30 m, maximum	
		depth not exceeding	
		2.00 m.	

During the process of taking off, descriptions are often written in a shorthand. The contents of the description are normally established with reference to the standard method of measurement. An extract from the concrete section of the SMM7 has been reproduced in Fig. 1.

E In situ concrete/Large precast concrete

E05 In situ concrete constructing generally
E10 Mixing/casting/curing in situ concrete

INFORMATION PROVIDED				

P1   The following information is shown either on location drawings under A Preliminaries/General conditions or on
  further drawings which accompany the bills of quantities:
  (a) The relative positions of concrete members
  (b) The size of members
  (c) The thickness of slabs
  (d) The permissible loads in relation to casting times				

Classification Table				
1 Foundations 2 Ground beams 3 Isolated foundations			m^3	1 Reinforced 2 Reinforced > 5% 3 Sloping $\leq$ 15° 4 Sloping $\geq$ 15° 5 Poured on or against earth or unblinded hardcore
4 Beds 5 Slabs 6 Coffered and troughed slabs 7 Walls 8 Filling hollow walls	1 Thickness $\leq$ 150 mm 2 Thickness 150–450 mm 3 Thickness > 450 mm			
9 Beams 10 Beam casings	1 Isolated 2 Isolated deep 3 Attached deep			1 Reinforced 2 Reinforced > 5%

Fig. 1

A heading would be required identifying the kind and quality of the concrete, any tests of materials and finished work, measures to achieve water-tightness, limitations on the method, sequence, speed or size of pouring and any methods of compaction and curing that might be specified. This can be found from the additional specification rules S1–S6 on the top right hand side of the SMM7 page.

Typically a description for a specific item will be built up as follows:

- Location of the work: from the first column of the classification table.
- Further dimensions or descriptions required from columns 2 and 3 of the classification table.
- The dimension must be measured using the units shown in the fourth narrow column, while any further information can be taken from the fifth column, where applicable.

Further columns give the measurement rules for each item, with clarification being made in the columns of definition rules and coverage rules. The ability to read and interpret the standard method is essential so that the rules can be applied to the measurement.

It is important to understand that you cannot put thickness $\leq$ 150 mm from the second column with foundations from the first column as a horizontal line differentiates between groups of items.

A typical description of concrete in a suspended floor slab would be:

			Sample descriptions
4.00 9.00 <u>0.20</u>	In situ concrete, mix 35 N/mm ^{2}, 20mm aggregate (S1) Slabs, thickness 150–450 mm, reinforced	This item is measured in cubic metres therefore 3 dimensions are required in the dimension column.

If two or more measurements are related to one description it is normal practice for them to be bracketed together as follows:

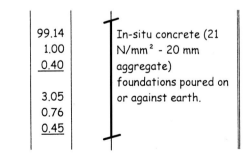

| | 99.14
1.00
<u>0.40</u>

3.05
0.76
<u>0.45</u> | In-situ concrete (21 N/mm ^{2} - 20 mm aggregate) foundations poured on or against earth. | |

The bracket is placed on the outside of the squaring column. A clear indication is made of where the bracket ends, the bracket itself usually being a vertical line with a short cross mark to indicate top and bottom. The examples in the subsequent chapters are considered to be sufficiently clear without the need for the use of brackets.

Anding-on

Where two or more descriptions are to be applied to one measurement, they are written as follows:

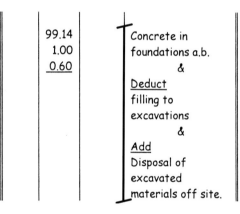

	99.14		Concrete in
	1.00		foundations a.b.
	<u>0.60</u>		&
			<u>Deduct</u>
			filling to
			excavations
			&
			<u>Add</u>
			Disposal of
			excavated
			materials off site.

In this case each description is separated by '&' on a line by itself. Care must be taken when 'anding on' in this way a superficial item with a linear item. It sometimes happens that it is very convenient to do so, but the distinction must be made quite clear so that the linear quantity is not used instead of the superficial. For example:

		<u>5.33</u>	5.33	25 x 100 mm wrot
				softwood rounded
		<u>5.79</u>	5.79	skirting plugged
	2/			&
		<u>7.00</u>	14.00	<u>Deduct</u>
				Emulsion general
		<u>4.50</u>	4.50	surfaces, internally
			<u> </u>	
			29.62	<u>x 0.10 = 2.96 m²</u>

To measure the deduction of the emulsion paint in this way saves setting down all the dimensions again as superficial items (exactly the same lengths being used). Similarly, a superficial item might be marked to be multiplied by a third dimension to make a cube, for example:

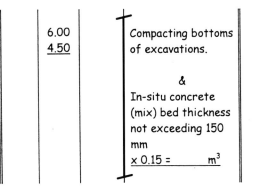

This approach is particularly useful for the measurement of roads, for example where you may have a large number of complex dimensions all relating to the various items associated with surfacing.

Deductions

Where a deduction is to be made, the description is preceded by *Deduct* which is often abbreviated to *Ddt* but when carelessly written has been known to be mistaken for *Add*. It is, of course, important to make quite clear whether a measurement is to be added or deducted and some surveyors always put the word 'Add' for any description immediately following a deduction, others only when a following addition is coupled to the deduction by '&', as in the example given below. In taking-off on traditional paper it is important that all deductions in a series of coupled descriptions are clearly marked 'Deduct', all doubt whether any description is an add or deduct thereby being removed. To add emphasis to the words 'Deduct' and 'Add' used in this context they are often underlined. With cut and shuffle taking-off it is necessary to keep deductions on separate slips, with the fact that they are deductions being clearly indicated.

An example of deductions follows:

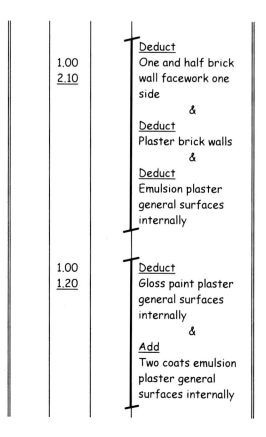

	1.00	Deduct	
	2.10	One and half brick wall facework one side	
		&	
		Deduct Plaster brick walls	
		&	
		Deduct Emulsion plaster general surfaces internally	
	1.00	Deduct	
	1.20	Gloss paint plaster general surfaces internally	
		&	
		Add Two coats emulsion plaster general surfaces internally	

Spacing of dimensions

One of the commonest faults found with beginners' taking-off is the crowded state of their dimensions. All measurements and descriptions should be spaced well apart, so that it is quite clear where one begins and the other ends. It is not unusual for a taker-off to realise, after writing down the measurements, that some item has been overlooked and should be inserted in its proper place. If the dimensions are well spaced out, it can be squeezed in, but otherwise it will have to be inserted elsewhere and cross-references made which only complicate the work. The use of a few extra sheets of paper will be found well worth while.

Accuracy

There may be a temptation, especially to those not sure of their ground, to take measurements on the full side, and so cover themselves against possible claims for deficiency. Such practice is to be deprecated, as it will be realised that, if measurements were on an average $2\frac{1}{2}\%$, or perhaps 5%, overmeasured, a very substantial amount would be added to the tender. Where figured dimensions are the basis the measurements taken should be exact, but where they are scaled there is some excuse for measurements being slightly full, as there may be some shrinkage of the paper and the lines on the drawing are not usually drawn more precisely to scale than, say, 50 mm on a 1:100 scale plan with any certainty. It will be seen, therefore, that the care with which the architect's drawings are prepared may affect the quantities, and that measurement from fully figured drawings must necessarily be more accurate. Furthermore, the quantity surveyor owes a duty to the client to prepare an accurate bill of quantities and it could be regarded as negligent not only to undermeasure but also to overmeasure.

When working from figured drawings, it may be found that three places of decimals are in some cases indicated. In setting down the dimensions the figures should normally be taken to the nearest two places of decimals, the excess or loss in measurement thereby obtained being usually very slight in proportion to the whole, particularly when the total bill quantity is usually given to the nearest metre. Three places may, however, be desirable when setting down waste calculations.

In short, the surveyor must aim at giving in the bill a representation as accurate as possible under the circumstances of the building work in question. There is, however, a sensible limit to the degree of accuracy to which one should measure; more experienced takers-off often take the value of the item into account when deciding on the degree of accuracy required.

Numbering the dimension sheets

All dimension sheets should be numbered as soon as possible. For those in traditional form, some surveyors number each column, others each page, this being a matter of individual preference. If the numbering is done at the earliest opportunity, it will minimise the risk of a page being mislaid and overlooked. If sheets are inserted after numbering, they can be numbered with the last number and a suffix a, b, c, etc., but it is essential that, if sheets numbered, say, 31a, 31b, 31c and 31d are inserted, a note should be put against the number 31 '31a–d follow'; otherwise the inserted sheets might be lost and not missed. A similar note should be made if gaps occur in numbering.

In any case, it will be found of value if the taker-off numbers the pages at the top, according to the section they represent, e.g. floors 1, roofs 6, plumbing 9, etc., irrespective of any other system of numbering. These pages can be kept in order awaiting the main sheet numbering, which may follow on from another section. It is further of value if on opening the dimensions at any page, one has an immediate indication of which section is measured. Special rules apply to the method of numbering of cut and shuffle sheets.

Cross-references

The taker-off should try to ensure that dimensions are clear to others, as it is quite possible that, when variations on the contract have to be adjusted, someone else will be entrusted with the work and will have to find their way around the dimensions. Cross-referencing between the drawings and the dimensions is essential for ease and speed in locating why particular dimensions have been used. It has been known for projects to be postponed for a year or two after tenders were received, and even the taker-off will then need some references and notes to refresh the memory. The dimensions may also need to be referred to when preparing the final account, and again clarity is important.

Clearness of the dimensions

Besides the use of cross-references, a good deal can be done to make the dimensions clear by the manner in which they are set down. It has already been pointed out that a regular order of length, width and depth (or height) should be maintained in writing down the dimensions; even when it may be difficult to determine which is length or width, a consistent order should be kept. In measuring areas of floor finishes, for instance, the dimensions horizontal on the plan could be put first, followed by the vertical ones. Calculations should be made as waste on the dimension paper and not mentally, and timesing should be done consistently. For instance, in measuring six doors each with four squares of glass, all timesed for two floors, the dimensions should be timesed as follows:

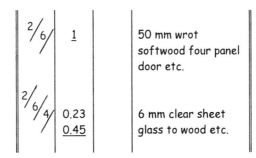

the last item not being written as:

When timesing becomes complicated, it will help considerably in tracing items if the method of timesing is consistent. In the example above, the outer timesing represents the floors and the next the number of doors; if the order is reversed in the middle of a series, there could be confusion. The use of coloured pens in timesing or dotting on to represent different floors or sections of the work will be found to help considerably in tracing dimensions later.

Headings

The use of headings in the dimensions will further aid future reference. Apart from the taking-off section heading already suggested for each page, subheadings or *signposts* should be clearly written wherever possible, and these will stand out if underlined. The sequence of measurement in the section will then easily be followed by a glance at the subheadings. Special subheadings will also be necessary where it is required by the SMM that a group of dimensions be billed together, as, for instance, in the detailed measurement of manholes. These bill headings should be written across the dimension and description columns and underlined to distinguish them from signposts, and the end of the work should be clearly defined by a note such as 'end of manholes'.

Notes

The making of notes by the taker-off on the dimensions is of the utmost value. Such notes are usually one of three kinds:

(1) Explanatory for use in writing the specification or adjusting variations
(2) For reference before the taking-off is finished, e.g. notes of items to measure or queries to be settled
(3) Instructions for preparing the bill.

If a specification is not available prior to taking-off, then the explanatory type of note should include memoranda for the specification, such as the rules followed in measurement of lintel reinforcement, or, say, in the case of a plate bolted down to brickwork, the spacing of bolts assumed. Such notes will save the specification writer from spending time trying to see on what principle reinforcement is measured or how far apart bolts were taken, and even if the taker-off writes the specification such notes will be of value. They will also be found of value in adjusting variations, especially if this is done by someone else. This type of note should be written at the side of the description column on waste, and is best separated by a line or bracket to prevent confusion with descriptions.

The second type of note, made for reference before the taking-off is finished, is necessary when perhaps some point must be referred to the architect or engineer, or for other reasons something cannot be finally measured. As mentioned previously, a list of such queries should be compiled. *To take* notes are entered in dimensions when a taker-off dealing with one section feels that an item, although arguably within the work being measured, is better taken with another section, e.g. tile splashbacks to sinks taken with the plumbing services rather than with the finishes or vice versa. These *to take* notes should be written clearly in the dimensions and collected together before the bill is prepared, and a check should be made to ensure that all items listed have been measured. Memoranda, too, should be made at the end of the day's work of anything unfinished which, the train of thought being broken, might be forgotten. It is much safer to make notes of such matters on the dimensions than to trust to memory, and these notes enable the work to be carried on in the case of unexpected absence. Such notes as these are sometimes written in pencil to be erased when dealt with, or they may be written in ink across the description column in such a way that they cannot be missed.

The third type of note is that which is an instruction to the preparer of the bill, used generally where a correction or alteration can be more easily made by a general instruction than by writing down new dimensions or descriptions. These notes, too, must be written in such a way that they will not be missed. An example of this type of note would be such an instruction as 'alter all 25 mm shelving to 32 mm'.

Insertion of items

It quite often happens that the taker-off must go back to the work and make alterations or insert additional items, either because items have been forgotton or because revised details have been received, as described above. Such alterations and additions should wherever possible be made in the proper place in the dimensions, so that when the dimensions are referred to at a later stage everything can be found where expected. Once more, the importance of plenty of space must be emphasised, and on work of any size it will be found of advantage to start each section of the taking-off on a new sheet, leaving any odd blank columns for later use if necessary. Further, if the dimensions are kept in subsections or groups with a definite gap between, these gaps will also be found of use for the insertion in their proper place of any dimensions as an afterthought. If it is impossible to insert an item or group of items in its proper place, a place must be found for it elsewhere and proper cross-references made in both places, as described previously.

Squaring the dimensions

A check is made of the waste calculations and their correct transfer to the dimensions. The dimensions are then calculated or *squared* and where bracketed together are totalled, subtracting any deductions that follow immediately. The *squarings* and casts are then checked and ticked.

Scheduling dimensions

The need for setting down dimensions on dimension paper in the traditional format can be quite time consuming for some items and it can be beneficial to use a schedule to collect together dimensions for repetitive items, such as steel beams or pipework. Take-off schedules may be prepared on A3 paper or a computer spreadsheet, identifying the item to be measured and referencing the schedule sheet. A location column is normal down the left hand side with perhaps varying sizes across the page. The right hand side could be used for fittings, accessories, girths of beams or other notes.

The small extract from a schedule shown in Fig. 2 has been used to measure concrete beams on an Excel spreadsheet and is ideal when a building is set out to a grid pattern and beams are repeated over a number of floors. A collection of lengths at the end of the schedule would then be transferred to the dimension paper or entered directly into a computer package. In Fig. 2, the length of a beam on grid line 1 A–E is 7.90 m. This is also repeated on grid line 2, therefore the figure is multiplied by 2 to give a total length of 15.80 m. Errors in working out waste calculations or extending dimensions are avoided when using a spreadsheet in this way. Care is still required to measure the lengths of the beams and to avoid any data entry errors.

Beams	Reinforced concrete beam sizes						
	400 × 500 mm		350 × 400 mm		350 × 300 mm		200 × 300 mm
1st floor							
1 & 2, A–E	2/7.90	15.800					
A & E 1–2	2/5.10	10.200					
A & E 2–3			2/2.975	5.950			
3 A & E				7.200			
A & E 3–6					2/8.825	17.650	
5 & 6, A–E					2/7.20	14.400	
c, 3–5						7.150	
4, B–E						5.650	
B, 3–5							7.150
Lengths		26.000		13.150		44.850	7.150

Fig. 2

Schedules are often used to measure steelwork, drain pipe runs and manholes, pipework, ductwork and cables, but any item that is very repetitive could be measured in this way. The examples used in this text book are not complex enough to warrant the use of schedules for measurement but further explanation of a drainage schedule has been included in Chapter 15.

Chapter 3
Alternative Systems

This book concentrates on the traditional method of taking-off in order to explain the basic principles which can then be applied to alternative measurement and billing systems.

This chapter briefly outlines the alternative systems that have been developed and adopted in order to achieve standardisation, increase the efficiency of bill production and eliminate the time-consuming data processing activities.

Computerisation of the measurement and bill preparation processes has been evolving over a period of time based on the use of standardised descriptions and advances in computer hardware and software.

Standardisation

The art of taking-off is to a large extent based on developing a systematic approach, having a sound knowledge of building construction, acquiring the mathematical skill to calculate and measure dimensions and the ability to write clear descriptions. Of these, probably the latter is the most difficult to master because it has to be built up from long experience, especially in estimating and dealing with errors and claims arising from inadequate descriptions. The taker-off has to ensure that descriptions comply with the requirements of the SMM and refer properly to British Standards, and often has to frame them under pressure of time from complex and sometimes incomplete drawings and specifications. It is no wonder, therefore, that occasionally errors and omissions occur in descriptions and it is not unknown for there to be diversity between different bills from the same office or even between different sections within the same bill, particularly in elemental formats. Furthermore, both personality and experience influence the taker-off's approach to compiling descriptions; and brevity, verbosity and English literacy all have their effect.

In former years an assistant became proficient in framing descriptions by experience gained in a working-up section preparing abstracts and bills. With the introduction of cut and shuffle and computerised systems, coupled with longer full-time education, this form of experience has been lost and alternatives have had to be established.

Standardisation was achieved either by compiling a standard bill or library within an individual organisation including all the descriptions likely to be encountered in the type of work usually dealt with or by using one of the published standard libraries.

Standard libraries

Most libraries rely on the fact that phrases within descriptions are frequently repeated whereas the complete description is often peculiar to one situation. Therefore, through splitting descriptions into phrases and allocating the phrases to various levels, full descriptions can be built up by selecting phrases at each level, some being essential and others optional. Considerable saving in bulk is achieved as phrases that are frequently used are listed once only.

The levels of phrases in a standard library could be as follows:

- Level 1: main work section
- Level 2: subsidiary classification
- Level 3: main specification in heading
- Level 4: description of item
- Level 5: size and number
- Levels 6 and 7: written short items (extra over items).

By selecting obligatory and optional headings and phrases from each level, combinations are built up to complete the description. The advantages of such a system are numerous; some of them are summarised in the following list:

- Descriptions are standardised and consistent.
- It provides an *aide-mémoire* for the taker-off.
- Billing can be carried out by less trained staff.
- Bill editing is greatly reduced.
- Consistency between bills aids price comparison of items.

- Simplest possible wording, almost in note form, avoids ambiguities.
- The use of ditto (often causing doubt) is avoided.
- It complies with the SMM.
- Consistency between bills from various sources aids the estimator.

This list would appear to give standard libraries an overwhelming advantage. There is, however, a danger that the taker-off will apply a readily available standard description to the item being measured even though it does not fit exactly, rather than compiling carefully a proper description, commonly called a *rogue item*. Furthermore, there is no doubt that if the taker-off has to keep referring to bulky documents the train of thought is interrupted and speed is reduced.

Computerised bill production

With rapid advances in computer technology, and as computer hardware and software applications have become comparatively less expensive, many firms have adopted one of the many computer systems now available. They have become widely used not only for use in bill production but also for assistance in carrying out most other quantity surveying functions. It is desirable that the operation of these other functions be carried out without having to re-enter the data stored for bill production, and therefore database systems are used.

A wide variety of Windows and DOS based bill of quantities production systems are now available, which have varying degrees of sophistication.

Although paper-based taking off can be manually entered into the systems, the majority now available have the facility to input dimensions directly by using either the keyboard or a digitiser. A digitiser is an electronically sensitive drawing board from which dimensions may be electronically scaled from the drawings into the system. Lengths, perimeters and areas of both simple and complex shapes are calculated automatically.

All systems are based on having a standard library of descriptions being built up from library text displayed on the screen. When a particular description is not available from the standard library, there is a facility to create rogue items. The number of rogues is likely to be more in a complex building project than in a simple one. It is therefore the ease with which creating and sorting rogues can be carried out that determines how efficient a particular system is.

An extract from a typical library of descriptions is shown in Fig. 3. This figure shows a sample of the various levels of the description that are available, along with a coding system. The main heading might only be used once in a bill of quantities but the common brickwork may well be required in both 102.5 mm and 215 mm thicknesses. By checking the requirements of SMM7 for brick and block walls the potential items required in a library can be extensive. Correctly coding items measured on paper for subsequent entry into a measurement package is of paramount importance, and there is always the danger of mis-coding. A record is kept, however, and can be physically checked.

Code	Description	Unit
F10	F10 BRICK / BLOCK WALLING	
F1001	COMMON BRICKWORK	
F1003	FACING BRICKWORK	
F10 - - 001	Walls	
F10 - - 003	Isolated piers	
F10 - - - - - 01	102.5 mm thick	m^2
F10 - - - - - 02	215 mm thick	m^2
F10 - - - - - 03	.... mm thick	m^2
F10 - - - - - - - 01	battering	
F10 - - - - - - - 02	tapering one side	
F10 - - - - - - - 03	tapering both sides	

Fig. 3

In contrast, some packages allow you to search through the library of descriptions to chose the item to include in your take off and it is less likely an error will be made at this stage. Dimensions can then be entered with the relevant description along with any waste calculations or side notes. A continual print of the data entry is useful so that you have a record of the logic used for the taking-off. This can prove very useful at post-contract stage when researching how the bill was measured so that variations can be assessed.

The order of taking-off for direct computer entry may differ from that used in the traditional approach. Utilising the 'anding-on' facility, or as some suppliers of software call it 'carry forward', can be extremely useful to save re-entering a long stream of dimensions. Before you start a section of measurement think carefully about how you want to order your work.

Common features with software packages are the facility to utilise data from previous bills of quantities, to calculate the weight of steel and reinforcement automatically and to display rules for calculating volumes. Systems can contain standard libraries for different standard methods of measurement and, being database systems, they have a multiple sort facility and can produce different bill formats as required. They can also sort, display and print data to different levels of detail. Draft bills can be produced for editing, and in many ways the editing process is more important than when using manual methods of bill production.

The development of bill production systems is ongoing and they are continually being upgraded. Many systems already have the facility to generate bill items directly from computer-aided design (CAD) data or other schedule data. As such fully computerised systems continue to be developed and become more widespread, the quantity surveyor will concentrate more on interpreting the data produced, ensuring their completeness and utilising them not only for tender documentation production but also for cost control and post-contract administration purposes.

Electronic data interchange (EDI) is also becoming more widely used: the bill is issued in electronic form to the contractors, who can then use it to price their tender for the work. The advantages of this are that it speeds up the transmission and receipt of information, reduces the paperwork involved and eliminates the wasteful repetition of rekeying information into a separate system (which is time consuming and error prone).

Cut and shuffle

The same general rules apply for setting down dimensions in a cut and shuffle system of taking-off as for the traditional method.

When using the cut and shuffle system the taking-off is done in exactly the same way as demonstrated in this book, except that each description is written in full and each description with its dimensions is written on a separate slip of paper. When the measurement is complete, the slips are sorted into bill order and the slips containing the same descriptions are brought together and totalled on one slip. Provided that care is taken in framing descriptions, and with some editing, the bill can be typed straight from the sorted slips. If the word 'shuffle' is taken to mean 'to mix at random' then 'cut and shuffle' is perhaps a misnomer and could be better called 'cut and sort' or, as it is sometimes termed, 'a slip sorting system'.

Cut and shuffle paper and rulings

A sheet of A4 size paper can be divided into three or four sections, which can be separated later by perforations or by cutting with a guillotine to form slips. The important thing is that the taker-off should write on a reasonably sized piece of paper, though for sorting purposes smaller pieces are more easily handled. Carbonated paper is often used so that a record of the original take-off can be kept. The sheets could also be photocopied prior to cutting to serve the same purpose.

Obviously, if every item is to be on a separate slip many slips will have only one or two dimensions on them. There will, however, be cases where there are so many dimensions for an item that they will need two or more slips. The first slip used for an item is called the *master*. This is the one from which the bill will be written or typed. Additional slips for the same item are knows as *slaves*. It will be noted that the collection of master and all the slaves (both adds and deducts) is required to arrive at the final quantity.

The normal dimension paper ruling needs some modification if the slips are to serve as a draft bill. The item number, description and calculated quantity must be together in a prominent place where the typist can find them without confusion with other matter on the sheet. Several *boxes* are therefore provided (see illustration below) and they are numbered for reference as follows:

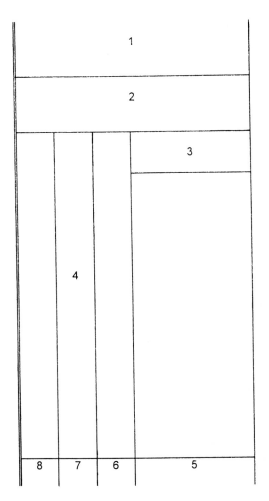

(1) This space is for the bill item number if the quantity surveyor wishes to number the bill items serially. These numbers will be entered immediately before typing the bill. If referencing of items is by serial letters on each page, this box will be left blank.

(2) This is for the description, which must be written carefully and in full by the taker-off on master slips, remembering that in doing this they are writing the final draft bill.

(3) When calculations are complete, squared and checked, this space on the master slip will have the quantity to be inserted in the bill with its unit of measurement; it will be left blank on slave slips.

These three boxes will be the only ones to be looked at by the typist and only the material to be entered in the bill must be put in them. If the description is a long one it can overflow into the top of the three dimension columns, but not into box (3).

The remaining numbers are:

(4) The normal dimension paper ruling, but the right-hand column is used only for waste calculations, location notes and location headings.

(5) Reference to the project number, taking-off section and slip number, e.g. 131/A/45 (project 131, substructure, slip no.). The numbering of each taking-off section should start at 1, the slips being pre-numbered or numbered as completed.

(6) Used for the total of the squared dimensions on the slip.

(7) In this box should be put C, S or L (cube, square or linear); it need only be used when the category is not obvious from the dimensions.

(8) This box can be used either for the work section letter as used in the SMM or for elemental references if preparing a bill of this kind.

Note: each slip is holed in the top left hand corner so that after cutting they can be filed in batches as convenient.

Rules for cut and shuffle taking-off

There are some rules that the taker-off must follow:

(1) Descriptions must be written in full as in a bill, with no abbreviations, except as in (5) and (6) below.

(2) Deductions must be on a separate slip and 'Deduct' must be written in box (2) and repeated in box (6) to appear against the total. This insertion in box (6) is very important; after completion of a section the taker-off, when looking through the dimensions, must check that all 'Deducts' are so inserted, to avoid mistakes occurring when the slips are processed.

(3) Two descriptions can be coupled to a single set of dimensions by '&' by writing the second description in full on the next slip

and by putting '&' in the dimension column. This is a case where C, S or L must be put in box (7).

(4) Cube items should be set down as cubes, not linear to be cubed up later.

(5) The word 'ditto' should never be used in descriptions, since when the sheets are cut there would be no indication of what is referred to. Instead, on slave slips a reference should be given to the master slip number, e.g. '25 × 10 mm s.w. skirting a.b. as 1/67', and abbreviations are then allowable. The description must be full enough to leave no doubt when sorting the separated dimensions: 'skirting a.b.' is not enough, as the sorter, who would not have 1/67 to hand, might think that it was tile, granolithic or something else.

(6) As many slave slips as necessary for each item may be used. Normal abbreviated descriptions can be used for such slips, provided that the item is clearly identified, remembering that the slips will be separated after cutting. Slave slips will eventually be attached to the master slip in such a way that the typist can turn over the whole batch, as the bill entry is made only from the master slip. Some surveyors prefer to detach the slave slips, leaving for the typist only those that have to be copied.

(7) If a whole slip is nilled, 'nil' should be written in the item number box (1) and the quantity box (3) as well as against the dimensions. It could also be written across any description.

(8) Slips that are left blank, e.g. at the end of a section to leave space for possible later additions, must have their serial number in box (5). They will be set aside in sorting.

(9) Specialists' work measured for a basis to obtain a PC sum should be measured together, so that these sheets can be taken out, cut and sorted in advance of the main bill.

Taking-off will proceed in the normal way apart from the restriction of a new slip for each new description. Dimension headings can be written as usual across the dimension columns. It will be an advantage to have standardised section and letter references for sections of the taking-off as below, and the taker-off must see that, before parting with the dimensions, they are marked in box (5) as described above.

Chapter 4

Applied Mensuration

Mathematical knowledge

It is assumed that the reader is acquainted with mensuration, knowledge of which, as of building construction, is an essential preliminary to the study of quantity surveying. It is rare that more detailed knowledge is required than the properties of the rectangle, triangle and circle. Less known formulae are detailed in Appendix 1 and can be looked up in an appropriate mathematical reference book. The properties of the rectangle, triangle and circle must, however, be thoroughly understood; if a student is not acquainted with them and with elementary trigonometry, these should be studied before going any further. This chapter shows some examples of how the theoretical knowledge of mensuration is applied to building work. Wherever possible, lengths should be found by calculation from figured dimensions on the drawings, rather than by scaling. Where scaling has to be used, a check should be made to ensure that other figured dimensions are accurate, as some reproduction methods affect the scale.

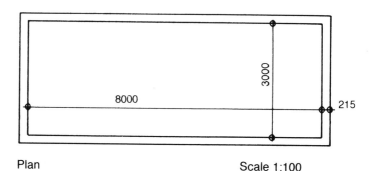

Plan   Scale 1:100

Fig. 4

Perimeter of buildings

Figured dimensions may be given internally or externally, and either can be worked from. Fig. 4 shows a plain rectangular building with one brick wall 215 mm thick.

	8.000
	3.000
	2/ 11.000
internal girth	22.000

The length of the perimeter of the external face of the walls, which may required for the measurement of external rendering, can be calculated by adding twice the thickness of the wall at each corner.

	22.000
4/2/ 0.215	1.720
external girth	23.720

This can be confirmed by adding the thickness of the wall to each internal dimension first.

3.000	8.000
2/ 0.215 0.430	0.430
3.430	8.430
	3.430
	2/ 11.860
	23.720

It will be seen that twice the thickness of the wall has been added for each corner.

Centre line of the wall

To arrive at the length of the wall as required by SMM7, the internal or external girth can be found and then an adjustment made to give the centre line as shown in Fig. 5.

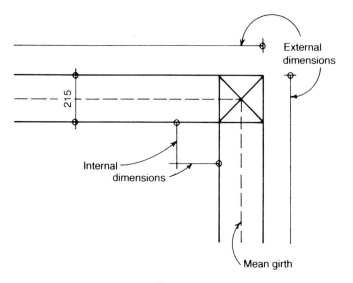

Fig. 5

A straightforward way of calculating the centre line measurement or mean girth from the internal dimensions is as follows:

$$
\begin{array}{ll}
\text{a.b.} & 22\,000 \\
\text{plus } 4/2/\tfrac{1}{2}/215 \;= & \underline{860} \\
& \underline{22\,860}
\end{array}
$$

It will be seen that twice of half the wall thickness is added to the internal perimeter for each external corner, i.e. 4 times, 2 times the distance moved (half the thickness of the wall).

The whole process can of course be reversed. If the external perimeter is taken instead of the internal, by deducting instead of adding, the mean length of the wall is obtained as follows:

$$
\begin{array}{ll}
 & 23\,720 \\
\text{Less } 4/2/\tfrac{1}{2}/215 \;= & \underline{860} \\
& \underline{22\,860}
\end{array}
$$

If the shape of the building is slightly more complicated, as shown in Fig. 6, the same principle may be applied to the calculations. For example the centre line or mean girth measurement is:

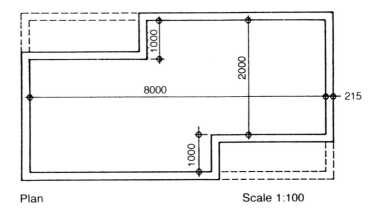

Plan   Scale 1:100

Fig. 6

2 000
1 000
3 000
8 000
2/11 000

Inside face   22 000
plus 4/2/½/215 =   860

Mean girth   22 860

Where the wall breaks back (enlarged in Fig. 7), the internal and external angles balance each other. As before, if the inside face has been measured, the external angle needs the thickness of the wall to be added to give the length on the centre line, whereas in the case of the internal angle the thickness of the wall must be deducted. In fact, the perimeter of the building is the same as if the corners were as shown dotted in Fig. 6. In short, to arrive at the mean girth the thickness of the wall must be added for every external angle in excess of the number of internal angles.

This collection of the perimeter of walls being of great importance, one further and more complicated example is given in Fig. 8. This time the calculations are made from the external figured dimensions instead of from the internal ones.

When smaller dimensions are given, you could check that their total equals the overall dimensions given, e.g. (see page 41):

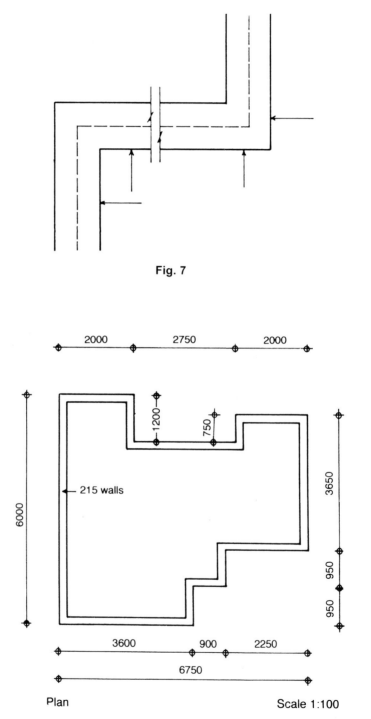

Fig. 7

Fig. 8

2 000	3 600		950
2 750	900		950
2 000	2 250		3 650
		1 200	5 500
6 750	6 750	− 750	+ 450
			6 000

Then the calculation of the mean length is as follows:

$$
\begin{array}{r}
6\,750 \\
6\,000 \\
\hline
2/\ 12\,750 \\
25\,500 \\
2/750 = \underline{\quad 1\,500} \\
27\,000 \\
\text{Less } 4/2/\tfrac{1}{2}/215 = \underline{\quad 860} \\
26\,140
\end{array}
$$

If you have difficulty in appreciating why 2/ 0.750 is added then colour the lengths of the wall that have already been measured carefully. This can make the extra length obvious. You could also think of this shape as having flexible joints and 'push' the shape out to form a rectangle. Again, you will be left with two additional lengths that will not 'fit', these being the 2/ 0.750 for the 'indent' marked a on Fig. 8.

The calculation of the mean girth is a most important one, as, having once been made, it is often used not only for several items in foundation measurement but also for brickwork and facings, with possibly copings, string courses, etc.

When calculating girths of the individual components of a cavity wall the same process should be followed. First the girth of the perimeter can be calculated, then progressively the mean girths of the outer skin, the cavity and the inner skin and then finally the internal face of the inner skin (as in Example 3 in Chapter 7).

Chapter 5
General Rules for Taking-off

In this chapter suggestions are given regarding procedure, general rules of measurement and other information that is necessary for the guidance of those responsible for coordinating the taking-off process, for the individual takers-off and for the study of the examples that follow.

It must not be forgotten that the methods of different individuals and the customs of different practices vary in the organisation and execution of this work. The fact that a particular suggestion is made here does not mean that it is universally adopted, nor does it preclude the use of an alternative.

Receipt of the drawings

As the drawings are received they should be stamped with the date and entered on a list showing the drawing number, title, date of preparation, date received and the number of copies. This list forms a useful reference and enables it to be seen when any revisions to previous drawings were received. A check should be made with the architect and engineer to ascertain what further drawings are under preparation which are likely to be received before the bill of quantities is completed. At the completion of the taking-off, each drawing used should be stamped 'used for quantities', which avoids confusion later on as to whether or not subsequent revisions were incorporated in the bill.

Preliminary study of drawings

Before any dimensions are written at all, the taker-off should look over the drawings and study the general character of the building.

A check should be made to ensure that all floor plans and elevations are shown and that the position of the sections are marked on all the plans. A more detailed inspection should be made to ensure that windows, doors, rainwater pipes, etc., shown on the plans are also shown on the elevations and vice versa. When drawings have been prepared by engineers they should be compared with those prepared by the architect to ensure that there are no discrepancies in the layout or dimensions. If overall dimensions are shown, these should be checked against the total of room and wall measurements. If overall dimensions are missing, these should be calculated and marked on the plans, and all projections on external walls should be dimensioned. If this is done the calculation of the mean girth and of the perimeter of the external walls is simplified. It is an advantage to dimension every room on the plans except where a series of rooms are obviously all of the same dimension in one direction. A little time spent in this preliminary figuring of the drawings obviates the possibility of inconsistency in dimensions. Where more than one taker-off is employed, the result of this work must be communicated to each of them, and their drawings marked accordingly. Figured dimensions must always be followed in preference to scaled and any dimensions that can be calculated from those figured should be worked out. A larger-scale drawing will usually override a smaller-scale one, except when the smaller is figured and the larger is not.

Queries with the designers

Preliminary inspection of the drawings may also give rise to a number of queries to be raised with the architect or engineer, relating to missing information or discrepancies. Settling questions at an early stage saves interruption to the taking-off and consequently increases productivity. It is important that all queries are set down and answered in writing. Queries should be listed on the left-hand side of a sheet of paper, with the answers, when received, shown opposite on the right-hand side, together with the date and source of the answer (Fig. 9). The quantity surveyor, however, must accept some responsibility in making decisions and many architects are quite ready to confirm such decisions. These query lists are then sent to the architect or the engineer for completion.

If materials are shown for which full up to date information is not available in the office it is advisable to ask manufacturers for

Project Nr	**Southtown School** **Queries**	Sheet Nr	
Date			
Ref *Query*		*Reply*	*Date*
1. Finish to floor of entrance hall specified wood block, coloured as tile?			
2. Should dimension between piers on north wall be 5.08 not 5.03? (to fit the overall 56.85)			
3. Dpc not mentioned in the spec. notes?			
4. Should brick facing to concrete beams be tied back?			
Etc. etc.			

Fig. 9

literature and advice at an early stage. If the survey of the existing site does not show levels or those that are shown are insufficient, it is prudent to request that a grid of levels is taken over the site. These levels will prove invaluable when measuring earthworks and are essential for reference later if there is any question of re-measurement.

Initial site visit

Before actually starting on the measurement it is advisable to make an initial visit to the site. If nothing else such a visit allows one to get the feel of the job and to be able to visualise later the various site references when they are encountered on the drawings. A further visit or visits will be necessary when the measuring has progressed in order to pick up the spot items (see Chapter 18) and site clearance items (see Chapter 17). The initial visit will give an indication of what will be required in due course; in this respect other items to be noted include such matters as boundaries generally (state and ownership of fences, walls, gates, etc.), the existence or otherwise of overhead and underground services, means of access, adjoining buildings (historic, civic, defence, etc.) and all matters that will be required for the subsequent drafting of the preliminaries bill (see Chapter 19). In addition, this visit is an opportunity to take the levels referred to above and also to examine and note any trial holes that have been dug and are open for inspection. A photographic record of the site or specific parts of it may prove useful as a reminder of existing site conditions.

Where to start

If all drawings are completed and available, the taker-off will probably follow the order of sections given below or such other order as may be the custom of the office. This order will be seen to follow more or less logically the order of erection of the building, except that certain special work and services are dealt with at the end. It may, however, be that the design of, say, reinforced concrete foundations is not completed, and therefore measurement cannot begin as usual with substructure. In such a case a point, such as damp proof course level, must be selected from which to measure the structure; the work below that level will be measured at a later stage when the necessary information is available. It may be that 1:100 scale drawings have been received, but 1:20 scale details are to follow. In such a case, internal finishes could probably be taken first, as the measurement of these is dependent generally upon the figured dimensions given on the 1:100 scale, which will probably not be altered. Moreover, these sections give the taker-off a good idea of the layout and general nature of the building. Specific forms of construction, e.g. a steel or concrete frame, will require a special sequence of measurement to be devised.

Organising the work

Where several takers-off are involved in the measurement of work and a deadline has to be met for the completion of the bill, it is important that the team leader organises the work carefully. To enable progress to be monitored, a schedule should be prepared showing the sections of work to be measured, with the name or initials of the taker-off responsible, together with the target and actual dates of commencement and completion of the work. Clear instructions must be given to each member of the group, carefully defining the extent of the work to be taken in each section. Proper arrangements should be made for the collection of queries for the architect or engineer and these should be edited by the team leader. Proper supervision should be made for junior staff involved and due allowance made for staff leave and their commitment to other work.

Sections of taking-off

The taking-off of dimensions is usually divided into sections under the main subdivisions of:

(a) Substructure (b) Superstructure (c) Finishes (d) Services (e) External works.

The sections found in a normal building would generally comprise the following:

(a)	Substructure	(1)	Substructures
(b)	Superstructure	(2)	Frame
		(3)	Upper floors
		(4)	Roof
		(5)	Stairs
		(6)	External walls
		(7)	Windows and external doors
		(8)	Internal walls and partitions
		(9)	Internal doors
(c)	Finishes	(10)	Wall finishes
		(11)	Floor finishes
		(12)	Ceiling finishes
(d)	Services	(13)	Sanitary appliances
		(14)	Disposal installations
		(15)	Water installations
		(16)	Heating installations/Air conditioning
		(17)	Electrical installations
		(18)	Gas installations
		(19)	Lift installations
		(20)	Communications installations
		(21)	Builder's work in connection with services
(e)	External works	(22)	Site works
		(23)	Drainage

A further subdivision for alterations and repair works, sometimes known as *spot items*, may be required.

The headings above cannot be regarded as mandatory, as there must be a certain amount of overlap. External facings and roof coverings might be considered as *finishes*, but, being part of the structure, they are usually included there. Floor finishings may be measured with the floor construction, but some surveyors prefer to take them with the internal finishes. The order of sections too may follow the particular preference of the surveyor, but it is best, having decided on a definite order, to stick to it. It must be understood that the special requirements of any building may require additional or subdivided sections; the list given above must therefore be regarded as flexible.

Taking-off by work sections

Some surveyors make a practice of taking-off by work sections instead of by elements of the building as described above. As the final bill is usually arranged in work sections, this system can minimise the sorting or abstract, which, when measuring by sections of the building, is necessary to collect and classify the items into work sections. Such a system may present difficulties in some cases, and would seem to increase the risk of overlapping or of forgetting something. When measuring by sections of the building, the taker-off mentally erects the building step by step and is less likely to miss items. Sometimes a combination of the two systems may be used to advantage.

Drawings

As already mentioned, figured dimensions on drawings should be used in preference to scaling. Naturally, where there is a large discrepancy between the two, one should compare with other dimensions given, or, if this fails to produce a solution, consult the architect. Care must be taken to use the correct scale when taking measurements from drawings, particularly where a variety of scales has been used. As items are measured from drawings it is often beneficial to loop through the applicable written notes and perhaps colour in the work on the drawing. An examination of the drawings marked up in this way, at the end of measurement, will soon reveal any items not measured. Not all measurable items are shown on drawings, particularly if the drawings are incomplete or in any case for labours such as surface finishes to concrete, but the method should prevent any major items being overlooked.

The specification

If a specification is supplied, it should be read through cursorily first, not with the idea of mastering it in detail, but rather with a view to getting a general idea of its structure and content. A more detailed study should then be carried out of the sections relating to the element to be measured. For example, the excavation, concrete work and brickwork should be studied closely before beginning the taking-off for the substructure. It is useful when the taking-off is well advanced to go through the specification and run through in pencil all parts that have been dealt with, but not paragraphs that will form preambles to the bill or that contain descriptions not repeated in the dimensions. If this is done it is unlikely that anything specified will be missed. Under the Joint Contracts Tribunal (JCT) form of contract, the specification is not a contract document; however, should the bill and specification both form part of the contract, they will have to be integrated together. In such cases the wording of the bill descriptions can be reduced by reference to the specification, particularly in a coordinated document. When the specification is a contract document, the preambles and the preliminaries in the bill can be similarly reduced in length.

More often than not, however, no formal specification is available at the outset. Brief specification notes may be supplied, and further notes must be made from verbal instructions given by the architect; these will form the nucleus of the specification, to be added to from time to time as queries are raised. These notes must be supplemented by the surveyor's own knowledge or by ideas of what is reasonably required. In making decisions on matters of specification the surveyor should bear in mind that what is theoretically correct is not necessarily the most practical or economic solution. In the case of timber, for instance, not only the difference between basic and finished sizes referred to in Chapter 8 should be borne in mind but also, where possible, the sizes that are readily available in the market should be specified. The proliferation of minor variations of the same article does not lend to economy, especially in joinery, where machines have to be reset for each different moulding. In fact the SMM in the case of door frames calls for the incidence of repetition of identical sets to be referred to.

Materials

The taker-off must have a through knowledge of the materials being used, and if a material is unknown should, where possible, make a point of seeing samples and studying the manufacturer's catalogue or leaflet, so that the limits and (perhaps optimistically stated) capabilities of the material are known. The handling of the material and study of any literature describing the materials and the way they are to be fixed in position will often assist in the measurement of the work or in the framing of a proper description.

Sequence of measurement

It is advisable when measuring to follow the same sequence in different parts of the work. For example, in collecting up the girth of the external walls of a building, it is a good idea to work clockwise, starting from, say, the top left-hand corner of the drawing. If this is done consistently it will assist in reference later, when perhaps the length of a particular section of wall has to be extracted from a long collection. If a particular sequence of rooms has been adopted in measuring ceilings, the same sequence should be used for wall finish, skirtings, floors, etc. In this way, if all the finishes of a particular room are to be traced, it will be known in which part of each group the relative dimensions are to be found.

Measurements for analysis of price

A distinction must sometimes be made between the measurements necessary for the preparation of a bill of quantities and those necessary for the analysis of a price for a particular item. This distinction is often not realised, especially by students. It has been found by experience that in certain cases a price can be built up for a unit of measurement without pressing the analysis of the item to its extreme limit, with all the additional work involved. For example, brickwork is measured including the mortar; the estimator who is accustomed to pricing this item will have an accurate idea of the rate for the whole item and, if not, the item can be analysed further to obtain its value.

Measurement of waste

It may be taken as a general rule (though like all rules this has exceptions) that measurements of work are made to ascertain the net quantities as fixed or erected in the finished building. Wastage of material generally is allowed for by the contractor in the prices, though sometimes a measurement is made as a guide to the amount of waste, and in a few cases the gross quantity is measured. The exceptions to the general rule will be pointed out as they occur.

Overall measurements

It is usual in measuring to ignore in the first instance openings, recesses and other features that can be dealt with by adjustment later. Brickwork, for instance, is measured as if there were no openings at all, and deductions are made when the windows, doors or other openings are dealt with in the proper section. Plastering and similar finishes are measured in the same way. It simplifies the work to measure in this way. When, for instance, windows are being considered the sizes will be to hand, and openings will be measured to correspond. Moreover, it may happen that, say, internal plastering and windows are being dealt with by different takers-off, when it is obvious that the one who measures the windows is better able to make the deductions and adjustments. Measuring overall with adjustment later is preferable to piecemeal measuring. This principle will also be found of value if, for example, a window should be forgotten, as the error would only involve the extra cost of the window over the wall and finishes, a much less serious matter than if nothing at all had been measured over the area of the window.

Use of schedules

Early preparation of information schedules of such items as finishes, windows and doors often exposes missing information and also provides a useful reference for the whole taking-off team. If like items are grouped together on the schedule the taking-off process becomes more straightforward and the need to continually refer to the specification during the taking-off is avoided.

Use of scales

A warning should be given of the possibility of using the wrong scale in measuring. If possible, each side of the scale used should not have more than one variety of marking on each edge, but this is not always practicable. The scale most easily available for general use is a standard metric scale having 1:5, 1:50, 1:10, 1:100, 1:20, 1:200, 1:250 and 1:2500 markings. Some surveyors prefer to have a separate scale for each variety, or one marked on one face only; but when working on two different drawings, e.g. 1:100 and 1:20, at the same time – as is quite common – it is a great convenience to have both markings on the same scale. In any case, special care is necessary when measuring from different drawings to ensure that readings are taken from the correct markings of the scale. It may seem unnecessary to emphasise this, but mistakes on this account are not unknown.

Use of SMM7

The forms of contract agreed by the JCT provide that measurements shall be made in accordance with the SMM and it is therefore of the utmost importance, where these forms of contract are used, that the standard method should be followed. It should, nevertheless, be understood that unless referred to in the contract, the SMM has no legal sanction and need not be adopted. However, this document has now been so long established that it could be produced as evidence of custom in the profession, and it is therefore advisable to study it and to follow its recommendations in all cases, or make quite clear where there is a divergence.

Decision on doubtful points

A thorough knowledge of the SMM, the measurement code (MC) and a study of all the published textbooks will still leave occasions when the taker-off must make decisions on the method of measurement or extent of descriptions for items not covered in the SMM. When a rule of measurement is originated in such a way it is often advisable to insert the method used as a preamble clause in the bill. When making such decisions the taker-off should have one main consideration: what will best enable the estimator to under-stand (not merely guess) the work involved and enable it to be priced quickly and accurately?

Descriptions

The framing of descriptions so that they are both clear and concise is an art not easily acquired, but one which is of the utmost value. The contractor's estimator, always working at high pressure, will waste much time if faced with long-winded and rambling descriptions, or having to decide what the surveyor intended to convey in a confused sentence. The surveyor, therefore, must always aim at clear expression, being careful in the choice of words and using the various technical terms in their proper sense. It is more common nowadays for standard description libraries to be used in which the descriptions are laid down for the taker-off. However, takers-off have to be very careful to avoid the pitfall of fitting an item to a standard description rather than ensuring that the description fits the item; if it does not then a special or rogue must be compiled.

The requirements of the SMM should be followed carefully when framing descriptions, but it must be remembered that addi-tional information should be given where necessary to convey the exact nature of the work to the estimator (SMM GR 1.1). Whilst it may be convenient to follow the order of the tabulated rules of the SMM, this does not prevent the use of traditional prose in the framing of descriptions (SMM Preface, second paragraph). Certain items such as waste of materials, square cutting, fitting or fixing materials or goods in position, plant and other items listed in Clause GR 4.6 of the SMM are generally deemed to be included in descriptions. Where the SMM calls for a dimensioned description to be given, apart from the description of the item, all dimensions should be given to enable the shape of the item to be identified (SMM GR 4.7).

The wording of descriptions is dealt with more fully in Chapter 19, but descriptions well drafted with the dimensions in the first instance will simplify considerably the work of the editor. The taker-off must also be careful to see that the same wording is used when referring to the same item in different parts of the dimensions, as inconsistency in the descriptions may indicate that some distinction must be intended by the different phraseology. For instance, if the taker-off, describing plaster writes '2 coat plaster on block walls' and then after several such items writes later '2 coat plaster on partitions' they may end up as different items in the bill although this was not the intention of the taker-off. Therefore, when the same item appears in different places it should be written in exactly the same form, or after the first time abbreviated with the letters 'a.b.' to indicate that exactly the same meaning as before is intended. It is important to confine the description to what is actually to appear in the bill, and not to add to it particulars of location or other notes that are merely for reference and not intended to go any further.

The descriptions written on the dimensions should be in the form that they are intended to appear in the bill, such parts that would normally be covered by a preamble (see Chapter 19) being omitted. Notes to assist in writing the preambles, particularly for the less common items, can be entered on the dimension sheet but should be kept well clear of dimensions or descriptions.

In some cases there will be added to the description of a superficial or linear item a note of the number included in the item so that the estimator can judge the average size of each, for example:

3/	0.22		4 mm clear sheet
	0.40		glass to wood with
			beads in panes not
			exceeding 0.15 m²
			(In Nr 3 panes).

Some of the requirements of the SMM for descriptions may be covered by general clauses or preambles to each work section. For example, SMM H71.S.4 says that the lap in joints of flashings shall be given. A preamble clause saying that all flashings are to be lapped 100 mm at joints would be sufficient and would probably save repetition.

Abbreviations

Under the traditional method of taking-off, abbreviations can be extensively used to shorten descriptions and individual words. However, in the case of cut and shuffle taking-off, such abbreviations are restricted to slave slip descriptions (see Chapter 2). In the examples that follow, abbreviations have been kept to a minimum.

A special note might perhaps be made here of the abbreviation 'a.b.' for 'as before'. Where this is used and might refer to more than one item it always refers to the last such item. For instance, a description '44 mm door a.b.' would refer to the last type of 44 mm door measured, if there had been several different varieties. If however, there is any doubt it is best to add to the brief description sufficient for it to be identified or to say '44 mm door a.b. col. 146', the reference to the column number being a definite guide.

Extra over

Some items are measured as *extra over* others, that is they are not to be priced at the full value of all their labour and materials, as these have to a certain extent already been measured. For example, fittings such as bends and junctions to drain pipes or angles and ends to gutters are measured as extra over. This means that the pipe or gutter is measured its full length over the fittings and the estimator when pricing the item assesses the extra cost for the fittings substituting the original item. On small pipes, although fittings are measured as extra over, the saving of pipe is so minimal that it will probably be ignored by the estimator. Such items are simply described as *extra for* where it is obvious what they are extra over, e.g. 'extra for bend' following immediately after an item of pipe obviously means 'extra over the cost of the pipe in question for a bend'. In the bill these items may be *written short* on the main item so that they are identified together. The measurement of an item as extra over something already measured as extra over should be avoided.

Dimensioned diagrams

There is a requirement in the SMM for a certain amount of drawn information to be made available. Most of this drawn information will be available from the drawings used for the taking-off and so the requirement only means ensuring that the information required by the SMM is included and that the requisite number of copies of the various drawings accompany the tender documents. However, in certain cases the SMM suggests dimensioned diagrams or sketches to elaborate a description. These diagrams or sketches can either be produced separately for inclusion in the text of the bill of quantities or drawn by the taker-off, with the dimensions to be processed as a separate operation.

These diagrams are defined in SMM GR 5.3. They will be either extracts from the drawings or drawings specially prepared for incorporation with the written description on the facing page. Alternatively the diagrams are collected together and printed on one or more sheets at the end of the bill, each diagram being given a number and being referred to by that number in the body of the bill. Diagrams should be marked 'plan', 'section', etc., and be drawn to scale, the scale being indicated or the dimensions figured.

Apart from the above, sketches will be made in the dimensions or on separate sheets to work out points of construction not detailed or to supplement the architect's details. If possible, these should be made on the dimension sheets where the particular work is measured, on a spare space on the drawings, or if made on separate sheets they should be carefully preserved to show at a later stage what the taker-off had assumed when measuring. Any details of importance should be confirmed with the architect.

PC items and provisional sums

For various reasons it is not always possible when quantities are being prepared to define finally everything necessary for the completion of the building. For instance, it may be necessary for the architect to select certain articles, such as sanitary appliances, ironmongery, etc., in consultation with the client, and the details of these may very well not have been considered at the early stage when tenders were being obtained. It is not unusual, therefore, to put in the bill *prime cost* (PC) sums for these items, which the estimator will include in the tender for goods to be obtained from a supplier, but which are subject to adjustment against the actual cost of the articles selected. The contractor is given the opportunity in the tender to add for profit to each of these items.

Furthermore, the architect may require certain specialist work to be carried out by specific firms. PC sums are included in the bill of quantities for such work, which the estimator will include in the tender. The firms employed for these works will be subcontractors of the general contractor, commonly known as *nominated sub-contractors*. The main contractor retains general control and responsibility. Again the main contractor is given the opportunity in the tender to add for profit and also for attendance as described below.

With regard to PC sums, the JCT forms of contract allow a 5% cash discount to the main contractor for nominated suppliers, but no discount is given for nominated subcontractors. Therefore it is important to keep these quite distinct.

Provisional sums are sums included for general contingencies and work to be done by the general contractor for which there is insufficient information for them to be adequately described in the bill of quantities. How PC and provisional sums are incorporated in a bill of quantities is fully dealt with in Chapter 18.

Attendance

An item of general attendance on nominated subcontractors is required by SMM A42.1.16 to be given in the preliminaries bill. This is a general item covering all subcontractors and the meaning of general attendance is defined in coverage rule C3 of SMM A42. It covers such items as use of erected scaffolding, temporary water and lighting supplies, messing facilities, etc., and items of a general nature applicable to all subcontractors.

Other or special attendance referred to in SMM A51.1.3 will differ according to the nature of the work and must be considered for each item. Examples contained in this rule include hardstanding, power, storage, etc. The conditions accompanying the quotation (often small print on the back) can be helpful in drafting this item.

Approximate quantities

Work to be carried out by the general contractor, which may be uncertain in extent, can also be provided for by means of approximate quantities, i.e. by measuring work in the normal way but keeping it separate in the bill and marking it 'approximate'. For instance, the foundations of a building, where the nature of the soil is uncertain, may be measured as shown on the drawings and additional excavation, brickwork, etc., measured separately and marked 'approximate' to cover any extra depth to which it may be necessary to take the foundations. It is thereby made clear that adjustment of these quantities on completion of the work is anticipated. Alternatively, if there is a considerable amount of such work it may be contained in a bill of approximate quantities. As noted above, if the work cannot be described adequately it is to be included as a provisional sum (SMM GR 10.2).

Quick reminder

(1) Prepare a drawing register and log any revisions received during bill preparation.
(2) Check the plans, sections and elevations for any clashes of information or missing details.
(3) Put headings onto the dimension paper and number sheets.
(4) Measure work from drawings in a set, logical pattern. One method frequently adopted is to start in the top left hand corner of the drawing and work clockwise.
(5) Write clearly and legibly and space your work out.
(6) Start by producing a query sheet and then a taking-off list for each section.
(7) Measure items in the same sequence as your taking-off list, using figured dimensions in preference to scaling from the drawings.
(8) Waste calculations should appear before descriptions.
(9) Give locational notes against your dimensions so that work can be followed easily.
(10) Use schedules where possible to save repetition.
(11) Use 'to take notes' when you do not have sufficient information to measure an item, but make sure to check these prior to final bill production.
(12) When your measure is complete go back over drawings and line through all of the items measured. Then repeat this process with your take-off list to ensure that you have not missed an item.
(13) Check balances to ensure, for example, that all excavated material has been disposed of or used as filling where appropriate.
(14) Stamp drawings as used for bill preparation.

Chapter 6
Substructures

Particulars of the site

Before beginning the measurement of substructures the drawings must be examined to ascertain whether the existing ground levels are shown in sufficient detail for calculating average depths of excavation. If the levels are not shown or are insufficient, then it is necessary to take a grid of the levels over the site. Irrespective of taking levels the surveyor should always visit the site to ascertain the nature and location of existing buildings, and details for preliminaries items and for the measurement of excavation work. Among items to be noted for the latter are vegetation to be cleared, the existence of topsoil or turf to be preserved, pavings or existing structures in the ground to be broken up and, if trial holes have been dug, the nature of the ground and the ground water level. Where the proposed work consists mainly of alterations, an early visit to the site will be essential, and most of the taking-off may even have to be carried out there. The measurement of alterations or *spot items* and the methods of dealing with this class of work are dealt with in Chapter 17.

Bulking

When measuring excavation, disposal and filling, the dimensions are taken as in the ground, trenches being measured along their centre line, multiplied by their width and depth as shown on the drawing. Soil increases in bulk when it is excavated but no account is taken for this in the bill of quantities, the estimator having to make the due allowance.

Removing topsoil

Where new buildings are to be erected on natural ground, it is necessary to measure a separate superficial item for the stripping of the vegetable soil or topsoil where it is to be preserved. This is measured over the area of the whole building including the projection of concrete foundations beyond external walls. A separate cubic item has to be taken for the disposal of the topsoil, giving the location. Any further excavation for trenches, basement, etc., would then be measured from the underside of such topsoil excavation. If there are existing paths, paving, etc. over portions of the area to be stripped, an item must be taken for breaking out existing hard pavings, as a superficial item stating the thickness and the material and measuring the removal as a separate item. Breaking out may be taken as extra over the excavation.

Where the site is covered by existing buildings, the pulling down is dealt with as described in Chapter 17, no item for stripping topsoil being necessary. Demolition is usually taken to existing ground level; breaking out below ground level is measured as a cubic item which may be taken as extra over the excavation.

Where the site is covered with good turf which is worth preserving, an item should be taken for lifting it (SMM D 20.1.4.1) and a separate item for relaying any to be reused (SMM Q30.4).

Reduced level excavation

Where a site is sloping it is often more economical to set the ground-floor level so that one end of the site must be excavated into, i.e. the underside of part of the hardcore bed will be below the level of the ground after the topsoil is stripped. Where this is the case, a cubic measurement is made of the excavating to reduce levels necessary from the underside of the topsoil excavation already measured to the underside of the hardcore bed. The depth for this item must be averaged, and it will generally be found that excavation is only necessary over part of the site, the level of the remainder being made up with hardcore or other filling. In Fig. 10 it can be seen that the ground level must be reduced to a formation level of 44.00 (300 mm below the top of the floor slab). The contour of 44.00, known as the *cut and fill line*, is plotted on the plan as accurately as possible from the levels given; the area on the right-hand side of this contour is measured for excavation, that on the left-hand side for filling.

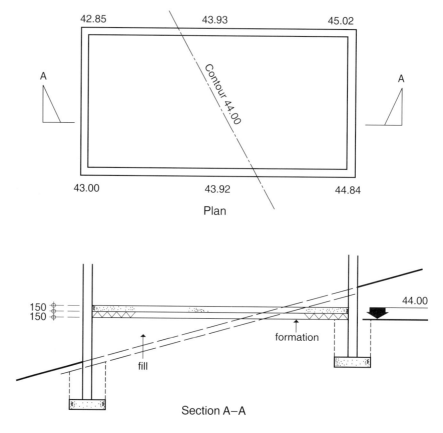

Fig. 10

Additional contours may have to be plotted to classify the excavation according to the maximum depths required by the SMM. It may, however, be more sensible to find an average depth for the whole excavation for measurement purposes and classify the depth in the description as the maximum on site. If this method is used a statement should be made in the bill to this effect.

The excavation to reduce levels must be measured before any trench excavation, as it brings the surface to the 'reduced or formation level' from which the trench excavation is measured, and, like the stripping of topsoil, it must be measured to the extreme projection of concrete foundations. A separate cubic item for disposal of excavated material must also be taken. It will often be necessary, where the floor level of the building is below the ground outside, to slope off the excavation away from the building, and possibly to have a space around the building excavated to below the floor level. In such circumstances the additional excavation would be measured with the external works section and a superficial item taken for trimming the side of the cutting.

Excavation for paths

Stripping of topsoil and excavation to reduce levels may also be required for formation of paths, paved areas, etc. External work of this nature is best measured all together after the building has been dealt with, as it usually forms a separate section in the bill. When paths abut the building (thus overlapping the projection of foundations), the whole excavation necessary for the erection of the building should be measured with the building, the extra width necessary for the paths only being measured with the paths.

Levels

Before foundations are measured, three sets of levels must be known:

(1)   Underside of concrete foundation
(2)   Existing ground level
(3)   Floor level.

(1) and (2) are necessary to measure trench excavation, and (1) and (3) are necessary to calculate correct heights of brickwork or other walling. The natural ground level, as has been pointed out, will probably vary and have to be averaged either for the whole building or for sections of it; if the floor level and the underside of the foundation are constant, the measurement of trench excavation is fairly simple. However, both the underside of the foundation and the floor level may vary in different parts of the building, there being steps at each break in level, and sometimes the measurement of foundations thus becomes very complicated. It will be found useful to mark on the plan the existing ground levels at the corners of the building – if necessary, these can be interpolated from given levels – and in the same way the levels of the underside of the foundation can be marked on the foundation plan (if any). If stepped foundations are required, it will prove helpful if the foundation plan is hatched with distinctive colours to represent the

varying depths of the underside of the foundations. If no foundation plan is supplied, the outlines of foundations can be superimposed on the plan of the lowest floor.

Trench excavation

The measurement of trench excavation and other foundation work
will normally divide itself into two sections:

(1) External walls
(2) Internal walls.

the former being dealt with first. In the simplest type of building a
calculation is made of the mean length of the trench as described in
Chapter 4. This mean length of the trench will also be the mean
length of the concrete foundations. The width of trench will be the
width of the concrete foundation as shown on the sections or
foundation plan. The depth of trench will, if the underside of
concrete foundation is at one level, be the difference between that
level and the average level of the ground, after making allowance
for stripping of the topsoil or excavation to reduce levels already
measured. Where the underside of concrete is at different levels,
theoretically the excavation for each section of foundation
between steps should be measured separately, the lengths when
measured being collected and checked with the total length
ascertained. However, it may be found in practice that, the step-
pings to bottom of the trench being small and the ground normally
falling in the same direction, an average depth can be determined
for larger sections of the building, if not for the whole. The depth of
trench excavation will have to be stated in the description in
accordance with the requirements of the SMM; if the depths vary,
it may be necessary to measure different parts separately to keep
to the SMM classifications. Alternatively, as mentioned above, in
excavation to reduce levels the maximum depth may be given and
a statement made in the bill giving the method used. The main
measurement of excavation for the external walls having been set
down, excavation for any projections on this foundation can be
measured.

   It has been assumed above that the external walls have trenches
of a uniform width all round. If there are several different widths,
as may be the case where the thickness of wall varies, each width is

dealt with separately, different lengths of the same width being collected together, and the whole being carefully checked with the total length ascertained. Trench excavation is measured as a cubic item and a separate cubic item of disposal must be taken. Trenches not exceeding 300 mm wide are kept separate (SMM D20.2.5).

Internal walls must be collected up in groups according to the width of their foundation and the average depths. Allowance should be made in the length for overlap where an internal wall abuts against an external wall, by deducting from the length of the internal wall the projection of concrete foundation to the external wall at this point. A similar allowance should be made where internal walls intersect. The necessity for this is best shown diagrammatically (Fig. 11).

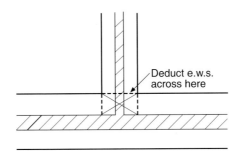

Fig. 11

It will be seen that if the foundation for the internal wall were measured the same length as the wall (i.e. as dotted in the figure) the area marked with a cross would be measured twice for the excavation and concrete. The amount involved being comparatively small, some surveyors may ignore the deduction, but to do so when there are many intersections is to take an unnecessary and definite over measurement without any justifiable reason.

Students often find it difficult to decide whether the maximum depths for trench excavation, etc. should be calculated from the original ground level or from the level of the ground after topsoil has been stripped, where this is measured separately. It is usual to measure from the latter level as this is the commencing level of the actual excavation being measured. The SMM requires that the commencing level of excavation must be stated where this exceeds 250 mm from the existing ground level, the depth classification of excavation being given from the commencing level.

Basement excavation

Basement excavation will be measured from the underside of the stripping of the topsoil or from the reduced level down to the underside of the basement floor, the trenches below being measured separately as trenches from the basement level. Where only part of the area of a building has a basement, it will be found most convenient first to measure all the basement complete up to a certain level, say, ground floor or general damp proof course and afterwards to measure the remaining foundations up to the same level.

Earthwork support

Earthwork support must be measured as a superficial item to the face of the excavation to be upheld, whether it will actually be necessary or not, to cover the contractor's responsibility to uphold the sides. It is for the contractor to decide, from the information on the nature of the soil, the extent and strength of support that will be required. Even if the decision is not to use any, there is still a risk to be priced; if the excavation falls in it will have to be re-excavated at the contractor's own expense. The surveyor measures the whole area to be supported, stating the distance between opposing faces as not exceeding 2 m, 2 to 4 m, or exceeding 4 m (SMM D20.7). The contractor may price the item at a nominal rate to cover a small risk, at the full value of complete support, or at some intermediate rate proportionate to the amount of support considered necessary. No extra excavation is measured to accommodate earthwork support, the contractor having to make due allowance for this. Earthwork support does not have to be measured for excavations less than 250 mm deep.

Where an internal trench intersects with an external trench then a deduction of earthwork support is required. This would be for the width and depth of the internal trench and is as shown previously on Fig. 11.

In Example 1, the earthwork support has been measured to the face of the surface strip, although that face does not exceed 250 mm, because the edge of the strip is in the same plane as the trench under and the total face to be supported exceeds 250 mm. Had the strip for any reason extended beyond the face of the trench excavation then support would not be measured.

Disposal of excavated material

It must be ascertained how the excavated material is to be disposed of. It is naturally cheaper if the material can be disposed of on the site, but there is often no room for it and it must then be removed from site. The special circumstances of each case must therefore be considered and the disposal fully described accordingly. Care must be taken to see that every item of excavation in the dimensions has an appropriate item of disposal measured. Where part is to be retained and part removed from site or otherwise disposed of, it will be found simplest to measure an additional item in the first instance as filling with material arising from the excavations equal to the volume of excavation as marked by the cross on Fig. 12. When concrete and brickwork are measured later an adjustment can be made of the volume occupied by these as:

Deduct Filling to excavations average thickness exceeding 0.25 m

&

Add Disposal of excavated material off-site.

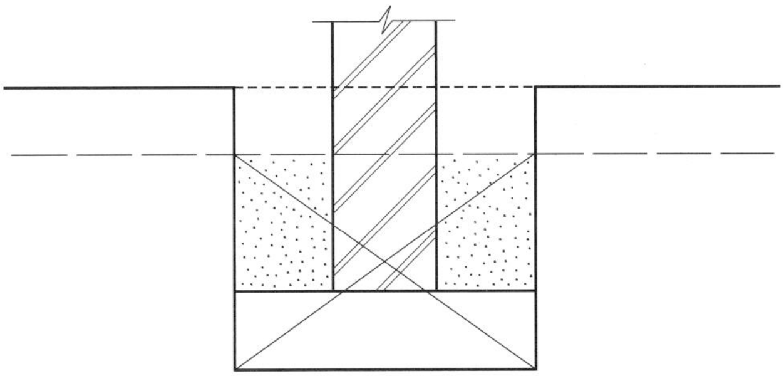

Fig. 12

If filling is deducted for the volume occupied by concrete and brickwork (shown hatched), that remaining of the original measurement will be the volume of space to be filled. It is simpler to calculate the volume of the brickwork than to obtain the volume of the spaces on each side. This is one example of the advantage of the overall system of measurement. Since all is measured in the first instance as filling, if adjustment for removal is forgotten, the error is less serious. Care is necessary in making the adjustment for the volume occupied by the wall to ensure that the height is not taken above the level to which the filling was measured in the first instance. In the case of basement excavation it is probably more convenient to measure all for removal, and adjust subsequently for filling round the outside.

If the filling round the walls is to be hardcore then this would be taken in the first instance, instead of filling with material arising from the excavations, together with a removal item for the excavated material. If only the inside of the trench is filled with hardcore then an adjustment will have to made for the filling to the outside, as demonstrated in Example 1.

Working space

A superficial item of working space allowance to excavations is taken for formwork, rendering, tanking or protective walls when it is necessary for workmen to operate from the outside and the space available is less than 600 mm. The measurement is taken as the girth or length of the formwork, etc. multiplied by the height measured from the commencing level of the excavation to the bottom of the formwork, etc. An area is measured, not a volume, as the estimator is left to make a judgement as to the width required giving due regard to the nature of the work involved.

Concrete foundations

The length measured for excavation of trenches will usually be found to serve for the centre line of the concrete measure. The width will be the full width of trench and the thickness as shown on the drawings. Where the concrete foundation is not at a uniform level throughout it will be found easiest to measure it as if it were, the additional concrete being added afterwards for the laps together with temporary support known as *formwork* to the face of steps. Again the advantage of overall measurement is apparent; if each section were measured piecemeal, there would be more danger of error through a section being missed. Concrete poured on or against earth or unblinded hardcore has to be so described.

Concrete foundations are sometimes reinforced by either steel fabric or bars. In such cases it is important to remember that a finer aggregate is necessary than in ordinary mass concrete, and the concrete will therefore be of a different composition but may otherwise be measured in the same way except that it is described as reinforced. A separate linear item is taken for the reinforcing bars, which is weighted up in the bill. Fabric reinforcement is measured as a superficial item, stating the width if in a one width strip.

When foundations are reinforced, weak concrete blinding 50 or 75 mm thick is usually required under the foundation. Such blinding is sometimes not shown on the drawings, in which case enquiry should be made of the architect or engineer as to whether it is wanted. Also the engineer may not be satisfied with earthwork support for forming the sides of concrete foundations and may require formwork, and if this is the case, provision for working space will have to be measured. Even if earthwork support is accepted, a flexible barrier or stabilisation of the excavated sides may be required.

Concrete ground slab

Hardcore or other filling to make up levels under concrete ground slabs is measured as a cubic item stating the average thickness as being either less than or greater than 250 mm. Compacting and blinding the surface of the filling is measured superficially, stating the material to be used. Damp proof membranes are also measured superficially as exceeding 300 mm in respect of the area in contact with the base, no deductions being made for voids less than $1.00\,m^2$.

Concrete ground floor slabs are measured as cubic items stating the thickness as not exceeding 150 mm, 150 to 450 mm; and exceeding 450 mm; the thickness excludes projections or recesses. When reinforced, this must be stated; slabs poured on or against earth or unblinded hardcore must be measured separately. Non-mechanical tamping of concrete surfaces is deemed included; however, other surface treatment such as power floating is measured separately.

Formwork is measured to the edges of beds and given linearly, stating the height in bands (SMM E20.2). Fabric reinforcement to slabs is measured in square metres.

Brickwork and blockwork in foundations

This work is often kept under a separate heading of *substructure*.

Where the concrete foundation is stepped, each length of wall from step to step will have to be measured separately up to damp proof course level. As this is piecemeal measurement the individual lengths taken should be totalled and compared with the mean girth of the entire wall. Alternatively, the total length may be measured by the minimum height, the extra heights being added for each section. When measuring internal walls, it should be noted that their length will not be equal to that taken for their concrete foundations as the length of these is adjusted at intersections as mentioned above.

Brickwork below damp proof course level is commonly specified to be in cement mortar, as opposed to gauged mortar for brickwork above, and this must be made clear in the descriptions. The wall between ground level and the damp proof course is often constructed of facing bricks, which may be continued for one or two courses below ground level to allow for irregularities in the surface of the ground. The measurement of walling in substructure follows the rules for the measurement of general walling, covered in the next chapter.

Damp proof courses

These are measured as superficial items and classified by width as exceeding or not exceeding 225 mm. Horizontal, raking, vertical and stepped damp proof courses are each kept separately. Pointing to the exposed edges is deemed to be included but the thickness of the material, the number of layers and the nature of the bedding material have to be stated. In the measurement no allowance is made for laps and no deduction is made for voids, such as flues, not exceeding $0.50\,m^2$. Asphalt damp proofing and tanking is measured using the area in contact with the base as a superficial item, and the widths are described as:

- Not exceeding 150 mm
- 150 to 225 mm
- 225 to 300 mm
- Exceeding 300 mm.

The thickness, number of coats, nature of base and any surface treatment have to be included in the description together with the pitch. Internal angle fillets are measured as linear items, stating a dimensioned description and the number of coats if other than two. Ends and angles are deemed to be included. Fair edges, rounded edges and arrises (or external angles) are measured linear unless the asphalt is subsequently covered when they are included in the description. Raking out joints of brickwork for a key is deemed to be included with the brickwork. No deduction is made for voids in asphalting not exceeding $1\,m^2$.

Example 1

Wall Foundations

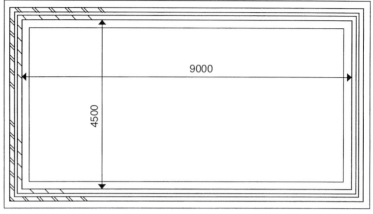

150mm topsoil to be preserved in spoil heaps

PLAN **Scale 1:100**

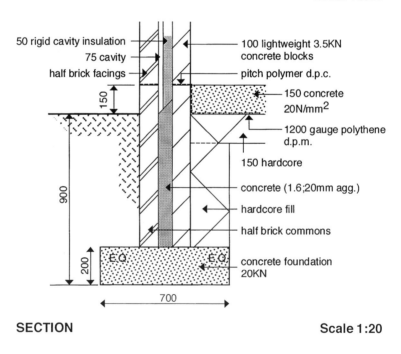

50 rigid cavity insulation

75 cavity

half brick facings

100 lightweight 3.5KN
concrete blocks

pitch polymer d.p.c.

150 concrete
$20N/mm^2$

1200 gauge polythene
d.p.m.

150 hardcore

concrete (1.6;20mm agg.)

hardcore fill

half brick commons

concrete foundation
20KN

SECTION **Scale 1:20**

Fig. 13

Example 1

Taking-off list	**SMM reference**
Topsoil excavation	D20.2.1.1
Topsoil disposal	D20.8.3.2
Trench excavation	D20.2.6.2
Disposal of soil	D20.8.3.1
Surface treatment	D20.13.2.1
Earthwork support	D20.7.1.1
Concrete foundation	E10.1.0.0.5
Cavity wall: brick	F10.1.1.1
block	
form cavity	F30.1.1.1
Cavity fill	E10.8.1
Adjust soil disposal	D20.9.2.3
Hardcore fill	D20.9.2.3
Topsoil backfill	D20.9.1.2.3
Damp proof course	F30.2.1.3
Adjust for facings	F10.1.1.1
Concrete bed	E10.4.1
Hardcore bed	D20.10.1.3
Surface treatment to hardcore	D20.30.2.2.1
Damp proof membrane	J40.1.1
Membrane upturn	J40.3.2
Concrete treatment	E41.2
Ground treatment	D20.13.2.3

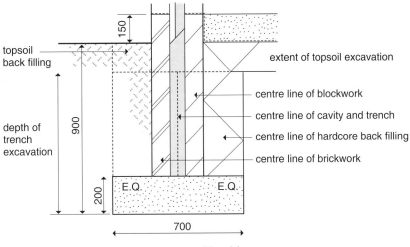

Fig. 14

				Substructure 1
			Half brick wall 102.5 Cavity 75 Block skin <u>100</u> <u>277.5</u>	Calculation of wall thickness
			Trench width 700 Less wall <u>277.5</u> 2)422.5 Foundation spread 211.25	
			<u>W</u> <u>B</u> 9.000 4.500 0.555 2/277.5 0.555 <u>0.422</u> 2/ 211 <u>0.422</u> <u>9.977</u> <u>5.477</u>	Calculation of overall dimensions to outer edge of concrete foundation
9.98 <u>5.48</u>			Excavating topsoil for preservation average 150mm thick & Disposal, excavated material on site in spoil heap average 100m from excavation x 0.15 = m^3 & Surface treatment compact bottom of excavation	D20.2.1.1 Only measured if required to be retained D20.8.3.2 D20.13.2.3 This ensures all compaction has been measured

				Substructure 2

Depth of trench

	0.900
less topsoil	0.150
	0.750

Mean girths

9 000	
4 500	
2/13 500	27.000
4/2/ ½ /277.5	1.110
Centre line of trench	28.110

	27.000
wall	278	
spread	211	
4/2/	489	3.912
Outer face trench		30.912

28.11	Excavating trench width	D20.2.6.2
0.70	exceeding 0.30m, maximum	
0.75	depth n.e. 1.00m	
	&	
	Disposal of excavated	D20.8.3.1
	material off site	This will be adjusted later
		for backfilling

				Substructure 3
2/	28.11 0.75 30.91 0.15		Earthwork support maximum depth n.e. 1.00m and n.e. 2.00m between faces	D20.7.1.1 Additional height of earth work support to topsoil excavation
				Trench depth 0.75m
	Item		Disposal of surface water	This shows how an item should be inserted into the dimension
	28.11 0.70 0.20		In situ concrete (20 N/mm²) foundation, poured against face of earth	E10.1.0.0.5

			Height of brickwork
			Depth of foundations 0.900
			To dpc 0.150
			                             1.050
			less concrete depth 0.200
			                             0.850	

In situ concrete (20 N/mm^2)

				Substructure 4
			<u>Girths</u>	
		½ / 102.5 51.25	27.000	These girths are taken from the internal girth as calculated before and using the 4 / 2 / distance moved principle.
		cavity 75		
		blocks <u>100</u>		
		4/ 2/ 226.25	<u>1.810</u>	
		c.l. brickwork	<u>28.810</u>	
			27.000	It is assumed all walls are vertical unless otherwise stated.
		4/ 2/ ½ / 100	<u>0.400</u>	
		c.l. blocks	<u>27.400</u>	
		½ / 75 37.5	27.000	
		<u>100</u>		
		4/ 2/ 137.5	<u>1.100</u>	
		c.l. cavity	<u>28.100</u>	
28.81 <u>0.85</u>		Walls, half brick thick in commons, in stretcher bond, in cement mortar (1:4)		F10.1.1.1 Outer skin brickwork up to dpc
27.40 <u>0.85</u>		Walls in dense concrete blocks 100mm thick in cement mortar (1:4)		F10.1.1.1 Inner skin blocks
28.10 <u>0.85</u>		Forming cavities in hollow wall, 75mm wide, including stainless steel twisted wall ties, 5 per square metre		F30.1.1.1 State width in description

				Substructure 5
			0.900	
		less foundation	0.200	Depth of cavity filling
			0.700	
		plus top splay	0.050	
			0.750	
28.10		In situ concrete (1:6 – 20mm		E10.8.1
0.08		aggregate) filling to hollow wall		
0.75		n.e. 150mm thick.		
		Trench depth	0.750	Depth of backfilling
		Less concrete	0.200	
			0.550	
		half spread		
		$\frac{1}{2}$ / 211.25 106	27.000	
		wall 278		
		4/2 / 384	3.072	
			30.072	
30.07		Deduct		D20.8.3.1
0.21		Disposal excavated material		This is the adjustment for
0.55		off site		earth filling to the outside of
				trench
		&		
		Add		
		Filling to excavations		D20.9.2.1
		exceeding 0.25m thick with		
		excavated material and		
		compact in 150mm thick layers.		
		Filling	27.000	
		Less 4/2 / $\frac{1}{2}$ / 211.25		
			0.845	
			26.155	

				Substructure 6
	26.16 0.21 <u>0.55</u>		Filling to excavations exceeding 0.25m thick with imported hardcore, compact in 150mm thick layers by vibrator.	D20.9.2.3 Inside filling
	30.07 0.21 <u>0.15</u>		Filling to excavations n.e. 0.25m thick with topsoil from site spoil heap n.e. 10m distance, lightly compact.	D20.9.1.2.3 Topsoil replacement outside buildings
	28.81 <u>0.10</u> 27.40 <u>0.10</u>		Damp proof courses n.e. 225mm wide, horizontal, pitch polymer, lapped 150mm at joints, bedded in gauged mortar (1:1:6)	F30.2.1.3
			<u>Facings</u> Three courses bwk 3/0.075 0.225	Taken one course below ground level and up to dpc
	28.81 <u>0.23</u>		<u>Deduct</u> Walls in commons as before & <u>Add</u> Walls, in facings, half brick thick, stretcher bond in gauged mortar (1:1:6) including pointing with a weathered struck joint as the work proceeds.	F10.1.1.1 The type of facing brick should be stated

				Substructure 7
			Ground floor slab	
9.00 4.50			In situ concrete (C20N/mm^2 – 20mm aggregate) bed n.e. 150mm thick x 0.15 = m3	E10.4.1
			&	
			Filling to make up levels n.e. 0.25m thick, imported hardcore compacted inlayers 150mm thick x 0.15 = m3	D20.10.1.3
			&	
			Surface treatment, compacting surface of filling and blinding with fine sand	D20.13.2.2.1
			&	
			Damp proof membrane exceeding 300mm wide, horizontal, 1200 gauge polythene laid on blinded hardcore to receive concrete	J40.1.1
27.00			Damp proof membrane width n.e. 200mm of 1200 gauge polythene sheeting at abutment	J40.3.2 Vertical turn up of damp proof membrane. Inside girth of blockwork, see note C2 in SMM 7

Example 2

Basement Foundations

(4m square internally on plan)

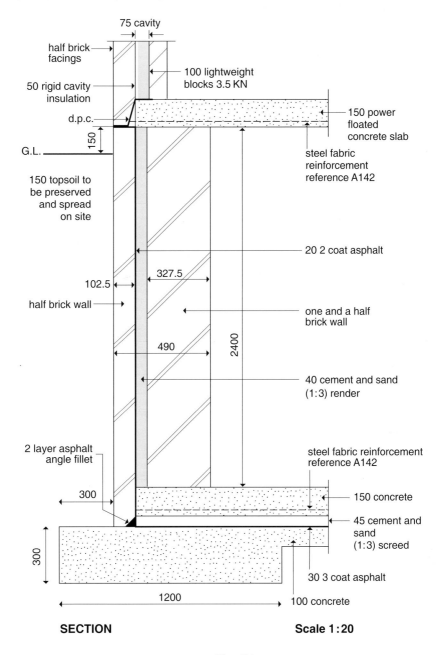

SECTION   **Scale 1:20**

Fig. 15

Example 2

Taking-off list	SMM reference
Topsoil excavation	D20.2.1.1
Topsoil disposal	D20.2.8.3.2
Surface treatment: bott. of exc.	D20.13.2.3
Basement excavation	D20.2.3.4
Disposal of soil	D20.8.3.1
Earthwork support: basement	D20.7.3.3
Trench excavation	D20.2.6.1.1
Disposal of soil	D20.8.3.1
Earthwork support: basement trench	D20.7.3.1
Working space: basement	D20.6.1
Trench foundation: concrete	E10.1.0.0.5
Concrete bed	E10.4.1.0.5
Surface treatment: concrete	E41.3
Asphalt tanking: horizontal	J20.1.4.1
Floor screed	M10.5.1.1
Concrete bed	E10.4.1.0.1
Mesh reinforcement	E30.4.1
Formwork to edge of bed	E20.2.1.2
Inner structural wall	F10.1.1.1
Wall render	M20.1.1.1
Asphalt tanking: vertical	J20.1.4.1
wall dpc	J20.1.2.1
angle fillet	J20.12.1
edges	J20.13
Half brick protective wall	F10.1.1.1
Concrete suspended slab	E10.5.1.0.1
Mesh reinforcement	E30.4.1
Surface treatment: concrete	E41.3
Formwork to slab: soffits	E20.8.1.1.2
edge	E20.3.1.2
Cavity tray dpc	F30.2.2.3.1
Adjust soil disposal: backfill	D20.9.2.1
remove from site	D20.8.3
topsoil	D20.9.1.2.3
Adjustment for facing bricks	F10.1.2

				Girths	Basement 1

		Girths	**Basement 1**
Wall 490		4 000	**4 m square internally on plan**
Spread 300			
2/790		1 580	As you become more familiar with calculating girths you can work out all the required girths at the start of your take-off and label for later use.
		4/5 580	
Outer girth		22 320	
- 4/2/ $\frac{1}{2}$ /300		1 200	
Backfill		21 120	
	-	1 200	Calculations of overall dimensions to outer edge of concrete.
Outer face		19 920	
-4/2/$\frac{1}{2}$ /102$\frac{1}{2}$		410	
		19 510	
	-	410	
		19 100	
-4/2/20		160	
Asphalt		18 940	
-4/2/40		320	
Render		18 620	
-4/2/$\frac{1}{2}$ /327$\frac{1}{2}$		1 310	
1 $\frac{1}{2}$ brick		17310	
	-	1 310	
Inner girth		16 000	
Outer girth		22 320	
-4/2/$\frac{1}{2}$ /1200		4 800	
trench girth		17 520	
		Depths	
		2 400	
Concrete 150			
Screed 45			
Asphalt 30			
Concrete 100		325	
		2 725	
less dpc – ground		150	
		2 575	
less topsoil		150	
		2 425	

				Basement 2
				Basement excavated to underside of concrete bed.
5.58 5.58			Excavating topsoil for preservation average 150 mm deep	D20.2.1.1 Required for re-use therefore measured
			&	
			Disposal excavated material, on site in spoil heap average 100 m from excavation x 0.15 = m^3	D20.8.3.2
			&	
			Surface treatment, compacting bottom of excavation	D20.13.2.3
			&	
			Excavating basements and the like maximum depth n.e. 4.00 m x 2.43 = m^3	D20.2.3.4
			&	
			Disposal excavated material off-site x 2.43 = m^3	D20.8.3.1

				Basement 3
			Depth of earthwork support 2 425 Topsoil 150 2 575	
22.32 2.58			Earthwork support n.e. 4.00 m deep, distance between opposing faces exceeding 4.00m	D20.7.3.3
17.52 1.20 0.20			Excavating trenches exceeding 300 mm wide n.e. 0.25 m deep commencing 2.58 m below ground level	D20.2.6.1.1
			&	
			Disposal excavated material off site	D20.8.3.1
			&	
			In situ concrete (20 N/mm² - 20 mm aggregate) foundation poured against face of earth	E10.1.0.0.5

				Basement 4
22.32 0.20			Earthwork support n.e. 4.00 m deep, distance between opposing faces n.e. 2.00 m	D20.7.3.1 Support not measured to inside face of trench as less than 0.25 m high

			Working space depth
			      2 425
			 plus topsoil 150
			      2 575
			 less slab 100
			      2 475
				Working space needs to be
				measured as a width of less
				than 600 mm free space is
				available

19.92 2.48			Working space allowance to basements and the like	D20.6.1
5.58 5.58 0.10			In situ concrete as before in bed n.e. 150mm thick poured against face of earth	E10.4.1.0.5

			Tanking
			     4 000
			Wall 2/327 ½ 655
			Render 2/ 40 80
			     4 735
			Asphalt 2/ 20 40
			     4 775	

				Basement 5
4.78 4.78			Trowel surface of concrete to receive asphalt	E41.3
			&	
			Mastic Asphalt Tanking/Damp proof membranes with limestone aggregate to BS 6925 finished with a wood float using fine sand as abrasive, subsequently covered.	J20 (S1 and S4) Work described as subsequently covered is deemed to include edges and arrises (C3)
			Tanking and damp proofing width exceeding 300 mm. 30 mm thick in three coats laid on concrete	J20.1.4.1 S2 & S3. Thickness, number of coats and nature of base to be stated
			Outer wall	

			Concrete 2 400
			               150
			Screed 45
			Asphalt 30
			             2 625
			Less asphalt dpc 20
			             2 605	

19.51 2.61			Walls in commons one and a· half bricks thick, vertical in stretcher bond in cement mortar (1:4) built against face of asphalt.	F10.1.1.1.1

				Basement 6
			Inner wall	
17.31 2.40			Wall, in commons one and a half brick thick, vertical in English bond in cement mortar (1:4) finish fair and flush pointed one side as the work proceeds	F10.1.2.1
18.62 2.40			Cement and sand (1:3) render to brick walls width exceeding 300 mm, 40 mm thick in three coats	M20.1.1.1
18.94 2.61			Asphalt tanking as before, Vertically, 20mm thick in two coats to brick	J20.1.4.1 The measurement is to the face of the wall. Rake out joints of brickwork for key is deemed included F10.C1(d)
19.51 0.10			Ditto level, width n.e. 150 mm 20 mm thick in two coats to brick	J20.1.1.1 To top of brick wall.
18.94			Solid external angle fillet to asphalt	J20.12.1 At junction of vertical and horizontal
19.92			Fair edge to asphalt	J20.13 Outer face of dpc

				Basement 7
	4.74		Sand and cement (1:3)	M10.5.1.1
	4.74		Screeds, floors level, 45 mm thick in one coat to asphalt base finished with wood float.	
			&	
			Reinforced in situ concrete (20N/mm2 – 20 mm aggregate) bed n.e. 150 mm thick	E10.4.1.0.1 lower bed
			$\times 0.15 = \quad m^3$	

$$\begin{array}{lr} \text{Inner face} & 4\,000 \\ \text{Plus wall } 2/490 & \underline{980} \\ & 4\,980 \\ \text{less } 2/\,102\,\tfrac{1}{2} & \underline{205} \\ & 4\,775 \\ \text{less asphalt } 2/20 & \underline{40} \\ & \underline{4\,735} \end{array}$$

4/	4.74		Formwork to edge of bed n.e. 0.25 m high	E20.2.1.2
			&	
			Formwork to edge of slab plain vertical n.e. 250 mm high	E20.3.1.2 Suspended high level slab.
	4.74		Reinforced in situ concrete slab n.e. 150 mm thick	E10.5.1.0.1
	4.74			
	0.15			

					Basement 8
				4 735	
			less 2/20 cover	40	
				4 695	
2/	4.70		Reinforcement fabric to BS 4483 type A142 weighing 2.22 kg/m² with 150 mm side and end laps		E30.4.1 for lower bed and suspended slab.
	4.70				

4 735
less 2/20 cover   40
4 695

2/   4.70
    4.70

Reinforcement fabric to BS 4483 type A142 weighing 2.22 kg/m² with 150 mm side and end laps

E30.4.1
for lower bed and suspended slab.

4.00
4.00

Trowelling on unset concrete bed to receive paving

E41.3

&

Formwork to soffits of slabs n.e. 200 mm thick horizontal 1.50 – 3.00 m high

E20.8.1.1.2
Upper slab

Dpc width
Block   100
Across cavity   75
Vertically   150
Brick   102
                427

Dpc girth
Face of brick   19 920
Wall $102\frac{1}{2}$
Cav   75
Blk   100
Less 4/2/ $\frac{1}{2}$ /277 $\frac{1}{2}$ 1 110
                 18 810

				Basement 9
	18.81 0.43		Damp proof course, exceeding 225 mm wide, horizontal cavity tray of pitch polymer bedded in gauged mortar (1:1:6)	F30.2.2.3.1
			Deduct	
	19.51 0.23		Half brick wall a.b	Facework F10.1.1.1.1
			&	
			Add	
			Half brick wall in facings a.b.	F10.1.2.1.1
			Deduct	Backfill
	21.12 0.30 2.33		Dispose excavated material off site	D20.8.3
			&	
			Add	
			Filling to excavations exceeding 0.25 m thick with excavated material and compacting in 150 mm layers	D20.9.2.1
	21.12 0.30 0.51		Filling to excavations n.e. 0.25 m thick with topsoil from spoil heaps average 100m distance and lightly compact	D20.9.1.2.3 Reuse topsoil from on site spoil heap

Chapter 7
Walls

Brickwork thickness

As the thickness of the brickwork has to be given in the description it is usually more convenient to give this in relation to the number of bricks. The generally accepted average size of a brick is $215 \times 102.5 \times 65$ mm, which with a 10 mm mortar joint becomes $225 \times 112.5 \times 75$ mm. This gives the following wall thicknesses:

- Half brick   102.5 mm
- One brick   215 mm
- One and a half brick 327.5 mm
- Two brick   440 mm.

Common and facing brickwork

When describing brickwork it is necessary to distinguish between common brickwork and facework. Common brickwork is walling in ordinary or *stock* bricks without any special finish and usually hidden from view. If the brickwork is exposed to view and finished with a neat or fair face then it is known as *facework*. Often facework is built with a superior type of brick which gives a more pleasing appearance. Walls one brick thick and over may have facework on one side and common bricks on the other, this being a more economical way of constructing the wall if only one face shows. Facework is described as to either one or both sides of the wall.

Measurement of brickwork

Brickwork generally is measured as a superficial item, taking the mean girth of the wall by the height. The heights of walls often vary within the same building but it is usually possible to measure to some general level, such as the main eaves, and then to add for gables etc. and deduct for lower eaves etc. No deductions are made for voids within the area of brickwork not exceeding $0.10 \, \text{m}^2$. In the case of internal walls and partitions, it is necessary to ascertain whether or not these go through floors and to measure the heights accordingly. If the thickness of a wall reduces, say at floor level or at a parapet, then it should be remembered that the mean girth of the wall changes. All openings in walls are usually ignored when measuring at this stage, the deduction for these being made when the doors or windows are measured, as described in later chapters.

Sub-division

The measurement of walls can be conveniently divided into the following subdivisions:

- External walls
- Internal walls and partitions
- Projections of piers and chimney breasts
- Chimney stacks.

The SMM requires that brickwork has to be described in the following classifications:

- Walls
- Isolated piers
- Isolated casings
- Chimney stacks.

In each of these classifications brickwork is described as vertical, battering, tapering (battered one side) or tapering (battered both sides). Isolated piers are defined as isolated walls when the length of the pier is less than four times its thickness, except where caused by openings.

Measurement of projections

Projections are defined as attached piers (if their length on plan is less than four times their thickness), plinths, oversailing courses and other similar items. Projections are measured linear and the width and depth of the projection are given in the description. Horizontal, raking and vertical projections are kept separately. Example 4 at the end of this chapter gives sample measures for these.

Descriptions

The descriptions for brickwork, in addition to the matters discussed above, have to include the kind, quality and size of the bricks, the bond, composition and mix of mortar and the type of pointing. These items, particularly if common to all the work, can be included in the bill as preambles or headings to reduce the length of descriptions. It should be remembered that, as already mentioned, mortar mixes may vary and, apart from substructure work, brickwork above eaves level in parapets and chimney stacks may well be specified to be in cement mortar.

Cutting and grooves, etc.

All rough cutting on common brickwork and fair cutting on facework is deemed to be included. Rough and fair grooves, throats, mortices, chases, rebates, holes, stops and mitres are all also deemed to be included.

Returns and reveals

The labour to returns at reveals at the ends of walls is deemed included and therefore nothing has to be measured for these.

Hollow walls

Each skin of a hollow wall is measured superficially and described as a wall, stating whether it has facework on one side. A superficial item is measured for forming the cavity, stating the width of the cavity. The type and spacing of the wall ties have to be given in the description. If rigid sheet insulation is required in the cavity this is also included in the description, stating the type, thickness and fixing method. Foam, fibre or bead cavity filling is measured separately as a superficial item, stating the width of the cavity, the type and quality of the material and the method of application. Filling cavities with concrete below ground level is measured as a cubic item stating the thickness in the description as not exceeding 150 mm, 150 to 450 mm, or exceeding 450 mm. Closing cavities is measured linear, stating the width of the cavity, method of closing and whether horizontal, raking or vertical.

Ornamental bands

These are items such as brick on edge or end bands, basket pattern bands, moulded or splayed plinth cappings, moulded string courses or moulded cornices etc. in facework. Horizontal, raking, vertical and curved bands are measured as separate linear items, stating the width and classified as follows:

- Flush
- Sunk (depth of set back stated)
- Projecting (depth of set forward stated)

If the bands are constructed entirely of stretchers or entirely of headers, this has to be stated, and if curved, the mean radius given. Ends and angles on the bands are deemed included.

Quoins

Labours to angles on brickwork are generally deemed included but when quoins, or corners, are formed with facings that differ in kind or size from the general facework they are measured linear on the vertical angle. They are classified in one of the three ways given for ornamental bands above. The method of jointing the quoins to the brickwork has to be stated together with the mean girth.

Copings

Copings in facings are measured linear and classified as for ornamental bands above. The description has to give the method of forming and the size. Sills, thresholds and steps are measured in the same way but are more likely to be encountered in the measurement of doors and windows. Tile or slate creasings are also measured linear, the width and number of courses being stated. Ends and angles are deemed to be included.

Reinforcement

Mesh reinforcement to brick joints is measured linear, stating the width and extent of laps. No allowance is made for laps in the measurement.

Measurement of arches

Where arches are to be measured or the wall below deducted, the measurement is fairly straightforward for semicircular arches. In the case of segmental arches the deduction above the springing line and the girth of the arch are not usually calculated precisely, as they can be estimated sufficiently accurately, the former from a triangle with compensating lines or by taking an average height, and the latter by stepping the girth round with dividers. In the case of expensive work, one should be as accurate as possible, the measurements preferably being worked out by calculation. A rough method of measuring the area of a segment is to take 11/16 times the area of the rectangle formed by the chord and the height of the segment. This is obviously not mathematically correct, and the margin of error will vary with the radius and length of chord, but when dealing with small areas this method will often be found sufficiently accurate. An alternative method is to take first the inscribed isosceles triangle based on the chord, leaving two much smaller segments, each of which may be scaled and set down as base $\times 2/3$ height. The error is then very small. In dealing with large areas of expensive materials a more accurate method would be necessary, using the formula:

$$\frac{H^3}{2C} + (\tfrac{2}{3} \, C \times H)$$

where C is the length of the chord and H the height.

Blockwork

The rules for measuring blockwork follow very closely those for measuring brickwork. Special blocks used, say, to achieve a designed bond, at reveals, intersections and angles are measured as linear items and described as extra over the work on which they occur.

Rendering

Rendering to external walls is usually taken with the walls and is measured the area in contact with the base as a superficial item. The description has to include the thickness, number of coats, mix and surface treatment, and the work has to be described as external. Rendering not exceeding 300 mm in width is measured as a linear item and work to isolated columns is kept separately. Rounded internal and external angles are measured as linear items. If metal beads are specified then these are measured linear and include the working of finish up to them. Painting to the walls is measured as a superficial item and described as to general surfaces externally.

Stonework

The measurement of stone walling is considered to be beyond the scope of this book but generally the rules follow those given for brickwork. Among the exceptions are isolated columns, rough and fair raking and circular cutting, grooves, throats, flutes, rebates and chases, all of which are measured linear.

Internal partitions

These partitions are mentioned with this section so as to complete the measurement of the internal walls although they are often measured with the internal finishes. When measuring timber stud partitions, all timbers including struts and noggins are measured linear and described as wall or partition members, stating their size in the description. All labours in the construction of the partition are deemed to be included. Door openings are usually measured net, as an adjustment later with the doors could be rather complicated. Plasterboard and other finish are measured as linear items, stating their height in 300 mm stages. Patent metal stud and other partitions are measured their mean length as linear items, stating their height in 300 mm stages, the thickness and whether boarded one side or both sides. All sole and head plates, studs, etc. are deemed to be included. Openings in the partitions are ignored unless they are full height in which case they are deducted from the length. Angles, tee junctions, crosses, fair ends and abutments to other materials are measured as linear items.

Cross-sectional dimensions of the timbers need to be given and the length must include an allowance for joints if mortice and tenon joints are to be used. The method of fixing and jointing should be included if it is not to be at the discretion of the contractor. One confusion that does arise is whether to include the wall, sole and head plates as partition members (G20.7) or plates (G20.8). It is generally accepted that all the members shown in Fig. 16 should be included with partition members.

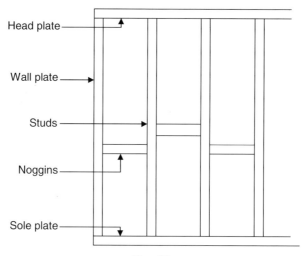

Fig. 16

Example 3

Walls and Partitions

(Above damp proof course)

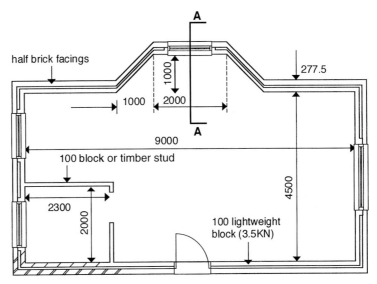

half brick facings

277.5

1000

1000 2000

9000

100 block or timber stud

2300

2000

100 lightweight block (3.5KN)

4500

75mm cavity with 50 rigid insulation

PLAN   **Scale 1:100**

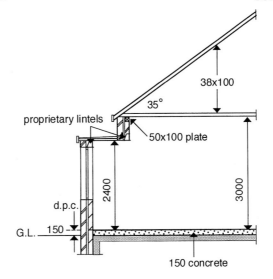

38x100

35°

proprietary lintels

50x100 plate

2400

3000

d.p.c.

G.L. 150

150 concrete

SECTION A-A   **Scale 1:100**

Fig. 17

Example 3

Taking-off list	SMM reference
Cavity walls: block	F10.1.1.1
facing brick	F10.1.2.1
cavity	F30.1.1
cavity insulation	P11.1.1
Adjustment for bay: last four items	
Lintel	F30.16.1.1
Gable ends	
Internal partitions	
Blocks	F10.1.1.1
Stud partition	G20.7.0.1

				External Walls 1

Calculation of mean girths excluding bay

The approach taken here is to find the internal girth and to use the 4 / 2/ the distance moved principle.

It is possible to adjust the internal girth to make allowance for the splay here but in order to explain the calculation in detail it has been taken separately on External Walls 3

	9 000	
	4 500	
2 /	13 500	
Internal girth	27 000	
4/2/ ½ /100	400	To centre of blocks
Int. skin	27 400	
	27 000	
4/2/ 137 ½	1 100	To centre of cavity
	28 100	
	27 000	
4/2/ 226 ¼	1 810	To centre of brick skin
	28 810	
	27 000	
4/2/ 125	1 000	To centre of insulation
	28 000	
	Height	See Fig. 13 for an explanation of the distances moved in each calculation
Floor – ceiling	3 000	
Less plate	50	
	2 950	

				External Walls 2
27.40 2.95		Walls, 100 mm lightweight concrete blocks (3.5 KN) vertical in gauged mortar (1:1:6)		F10.1.1.1
28.81 2.95		Walls, half brick in facings, vertical in stretcher bond in gauged mortar (1:1:6) including pointing with a weathered joint as the work proceeds one side.		F10.1.2.1 Type of facing would be given or a Prime Cost allowed
28.10 2.95		Forming cavities in hollow walls 75 mm wide including stainless steel twisted wall ties (five per square metre)		F30.1.1
28.00 2.95		50 mm cavity wall insulation including polypropylene insulation retainers (five per square metre)		P11.1.1

				External Walls 3
				$x = \sqrt{1.00^2 + 1.00^2}$
			Adjustment at bay	$= \sqrt{2.00}$
				$= 1.414$
			Internal face length difference	
			Bay 2 000	
			2/ 1414 2 828	
			4 828	
			Deduct	Length X remains the same for both walls. Length Y increases by the same amount as Z decreases
			2 000	
			2/1 000 2 000 4 000	
			Extra length 828	
0.83 2.40			Block walling 100 mm thick a.b.	This dimension could be added to the previous four measured items
			&	
			Wall in facings half brick a.b.	
			&	
			Forming cavities a.b.	
			&	
			Cavity insulation a.b.	

				External Walls 4

| | 1 | | Proprietary lintel type x
4 300 mm long and building
into blockwork | F30.16.1.1
This is lintel across bay.
Lintels over windows would be
taken with window measure |

		4 000
	Bearing 2/150	300
		4 300

| | 4.30 | | Deduct | F10 (M.3)
Full courses only deducted |
| | 0.23 | | | |

Block walling 100 mm thick
a.b.

ALTERNATIVE IF
BUILDING GABLED BOTH
SIDES

Blockwork

	4 500
2/100	200
Base of Gable	4 700
x ½ =	2 350
2 350 x tan 35°	
Height of gable =	1 646

Say 120 Top of rafter = Top of gable wall
Say 150   35°
Say
50   38 × 100
         150
         Height of
         main
         brickwork
         4500

extra hight
plate 50
rafter 100
       150

2/	4.70		100 mm block walling a.b.	These could be added to original item
2/½/	0.15			
	4.70			
	1.65			

					External Walls 5

Brickwork

		4 500
2/277½		555
		5 055
× ½ =		2 527½

2 527½ × tan 35°

Height =   1 770

2/	5.06	Half brick wall in facings a.b.
	0.05	
2/½/	5.06	
	1.77	

Cavity

		4 500
	100	
	75	
2/175		350
		4 850
× ½ =		2 425

2 425 × tan 35°

Height =   1 698

2/	4.85	Forming cavities in hollow
	0.12	walls a.b.
2/½/	4.85	
	1.70	

2/	4.80	50 mm cavity insulation a.b.
	0.14	
2/½/	4.80	
	1.68	

		4 500
	100	
	50	
2/150		300
		4 800
× ½ =		2 400

2 400 × tan 35°

Height =   1 681

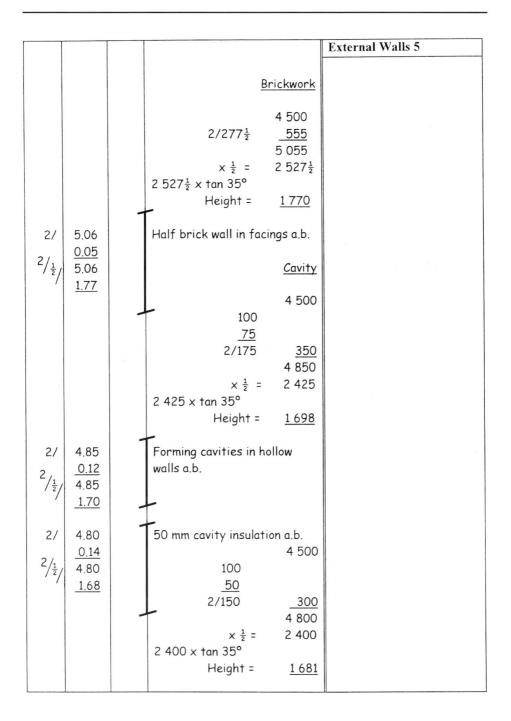

Alternative stud partition example

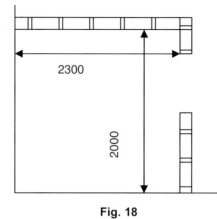

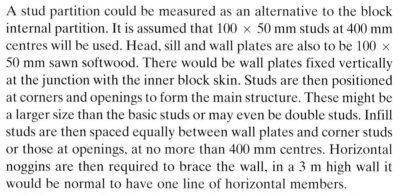

Fig. 18

A stud partition could be measured as an alternative to the block internal partition. It is assumed that 100×50 mm studs at 400 mm centres will be used. Head, sill and wall plates are also to be 100×50 mm sawn softwood. There would be wall plates fixed vertically at the junction with the inner block skin. Studs are then positioned at corners and openings to form the main structure. These might be a larger size than the basic studs or may even be double studs. Infill studs are then spaced equally between wall plates and corner studs or those at openings, at no more than 400 mm centres. Horizontal noggins are then required to brace the wall, in a 3 m high wall it would be normal to have one line of horizontal members.

Adjustments for the door opening should be made after the members have been measured. To decide how many studs are required, the length of the wall is taken and divided by the spacings. The result must be rounded up to the nearest whole number and then, as this is the number of spacings, it is necessary to add one to allow for the end stud.

				Internal Walls 1
			Lengths	
			2 300	
			2 000	
			4 300	
		2/ ½ / 100	100	
			4 400	
4.40			Walls in lightweight concrete	F10.1.1.1
3.00			blocks (3.5 KN), vertical in	
			gauged mortar (1:1:6)	
			ALTERNATIVE FOR STUD	See Fig. 18
			PARTITION	
				The studs are 100 x 50 mm at
			2 300	400 mm centres
			plus corner 100	
			400) 2 400	
			6 + 1	
			7	
			400) 2 000	
			5 + 1	
			6	An adjustment will
				subsequently be made for the
			+ 1 Extra at corner	door opening.
			total = 7+6+1 = 14	
			Height	
			3 000	
			less 2/ plates 2/50 100	
			2 900	

				Internal walls 2
2/	2.40		50 x 100 mm sawn treated softwood wall or partition members	G20.7.0.1 Head and sole plates
2/	2.00			
14/	2.90		noggins	Studs
	3.70		2 400 2 000 4 400 less studs 14/50 700 3 700	
	0.90		Door adjustment Taken as size 900 x 2100 high height for deduction 2 100 plus head 50 2 150 noggins width 900 less vertical stud 50 850	Door head
	2.15		Deduct last item	One vertical stud displaced by door and horizontal noggin
	0.85			

Example 4

Brick Projections

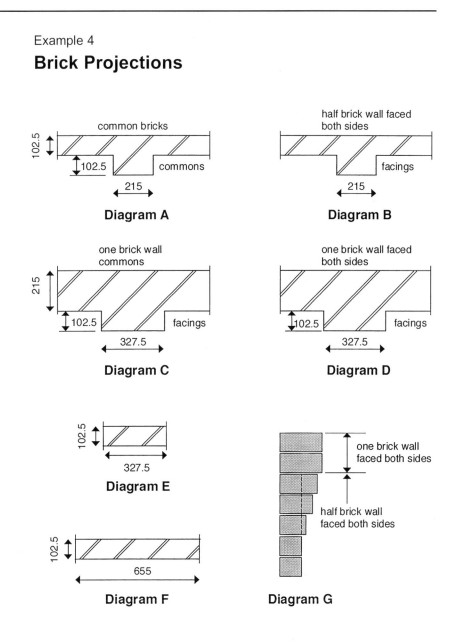

Diagram A

Diagram B

Diagram C

Diagram D

Diagram E

Diagram F

Diagram G

Plans: Diagrams A, B, C, D, E & F all 3000 high
Sections: Diagram G 5000 long, 3000 high

Scale 1:20

Fig. 19

Example 4

Taking-off list		**SMM reference**
Projections		
Diagram A		F10.5.1.1
Diagram B	Adjust main wall	
	Projection	F10.5.1.1
Diagram C	Adjust main wall	
	Projection	
Diagram D	Adjust main wall	
	Projection	
Diagram E	Isolated pier	F10.2.1.1
Diagram F	Wall	F10.5.1.1
Diagram G	Adjust main wall	F10.5.1.1
	Projection	F10.5.1.3

				Brickwork 1
			Note:-	
			In the following examples it has been assumed that the main wall has been measured ignoring the projection.	
			DIAGRAM A	
3.00			Projections in common bricks, one brick wide and half brick projection, vertical in gauged mortar (1:1:6).	F10.5.1.1 Measured as a projection if the length of the pier on plan does not exceed four times the projection.

				Brickwork 2
			DIAGRAM B	
0.22 3.00			Deduct Half brick wall in facings, vertical, pointed both sides & Add Ditto but pointed one side.	The original half brick wall in facings is substituted for the width of the pier by a wall pointed one side.
3.00			Projections in facing bricks, one brick wide and half brick projection, vertical, in gauged mortar (1:1:6) including pointing with a weathered joint as the work proceeds.	F10.5.1.1 F10.Cl(f) Labours on returns deemed included therefore no requirement to state returns in description.

					Brickwork 3

0.33 __3.00__			DIAGRAM C Deduct One brick wall in commons, vertical, faced and pointed one side. & Add One brick wall in commons, vertical, English Bond in gauged mortar (1:1:6).	The original wall faced one side is substituted with a one brick wall in commons for the width of the pier.
__3.00__			Projections in facing bricks, in English Bond one and a half brick wide and half brick projection, vertical, in gauged mortar (1:1:6) including pointing with a weathered joint as the work proceeds.	F10.5.1.1

				Brickwork 4
			DIAGRAM D	
0.33 3.00			Deduct One brick wall in facing bricks, vertical, pointed both sides a.b. & Add One brick wall in commons, vertical, English Bond in gauged mortar (1:1:6) faced and pointed one side with a weathered joint as the work proceeds.	The original one brick wall in facings is substituted with a one brick wall in common bricks faced one side for the width of the pier.
3.00			Projection in facing bricks in English Bond one and a half bricks wide and half brick projection, vertical, in gauged mortar (1:1:6) including pointing a.b.	F10.5.1.1

				Brickwork 5
			DIAGRAM E	
0.33			Isolated pier, half brick thick, vertical, in common bricks stretcher bond in gauged mortar (1:1:6).	F10.2.1.1 D8:- Taken as isolated pier because length is less than four times the thickness.
3.00				
			DIAGRAM F	
0.66			Walls, half brick thick, vertical, otherwise as last item.	F10.1.1.1 Length more than four times thickness therefore measured as a wall.
3.00				

				Brickwork 6
			DIAGRAM G	
			2/ 75 = 150 3/ 75 = 225 ——— 375	Calculation of heights of brick courses
			Deduct	
5.00 0.38			Walls half brick thick in facing bricks, vertical, pointed both sides a.b.	The original half brick wall has been measured for a height of 5 m and a length of 3 m, pointed both sides. This is now substituted with a wall pointed one side for the area of the oversailing courses
			Add	
5.00 0.23			Ditto but pointed one side	
5.00 0.15			Ditto one brick thick, pointed both sides a.b.	
5.00			Projections in facing bricks, average quarter brick thick and three courses high, horizontal in gauged mortar (1:1:6) including pointing with a weathered joint a.w.p	

Chapter 8
Floors

Timber sizes

Timber for use in construction is sawn into standard sectional dimensions and the size thus created is known as the *nominal* or *basic* size. When the timber is processed, planed or wrot it is reduced in size, usually by about 2 mm on each face, and the resultant size is known as the *finished* size. The original size of processed timber is sometimes described as *ex* or *out of*, thus 'ex 100 × 25 mm' would be '96 × 21 mm finished'.

Regularising is a machine process by which structural timber is sawn to a uniform size, e.g. joists sawn to an even depth. For regularising, 3 mm should be allowed off the size of timber of up to 150 mm, and 5 mm above this size.

The SMM states that sizes of timber are deemed to be nominal unless described as finished. Confusion may be avoided if the sizes given in the bill are as they are shown on the drawings, provided that these are consistent. Structural timber when not exposed is usually left with a sawn finish and the sizes given on the drawing are nominal. When architects are designing joinery, which may have to be to precise dimensions, they often prefer to give finished sizes. Care must be taken to ascertain which sizes are given on the drawing and to make it clear in the bill if finished sizes are being used.

The amount of timber removed in creating a wrot face on timber is known as the *planing margin*; this should be given as a preamble in the bill. Reference may be made to BS 4471 'Dimensions for Softwood', which sets out in detail various planing margins according to the sizes and the end use of the timber.

The specification for tongued and grooved floor boarding can cause confusion, as the finished width on face and the finished thickness may be quoted; e.g. 90 × 21 finished boarding would be

100×25 nominal (sawn) size to allow for the tongue and planing margin.

Further consideration should be given to the sizes of timber readily available. For example, the width of a door lining to a 75 mm block partition with 13 mm plaster on both faces would require a finished lining width of 101 mm. The nearest nominal size of timber available is 115 mm, or 125 mm for some types of timber.

Timber exceeding 5.1 m in length is usually more expensive; in the examples an allowance in the length of timbers has been made for jointing where appropriate. The SMM requires structural timbers over 6 m long in one continuous length to be measured as such, including stating the length (G20.6.0.1.1).

Subdivision

The measurement of floors is conveniently divided into two main subdivisions of finishes and construction. These subdivisions may be taken storey by storey or the finishes may be taken first for the whole building, followed by the construction. Whether or not the finishes are taken before the construction is really a matter of personal choice although measurement, and therefore knowledge, of the finishes may assist in deciding the form of construction to be used. For example, a timber upper floor may have a general finish of boarding but a small tiled area may require a different form of construction.

Frequently floor finishes are measured with the internal finishes section, because measurements used for ceilings may apply to floors. This is, however, a matter for agreement between the surveyors measuring the two sections, although it is customary to measure the finish to timber floors with the construction.

Timber floor construction

The measurement of timber floors may be divided as follows:

- Plates and/or beams
- Joists
- Hangers
- Strutting
- Ties to walls
- Insulation.

Timber suspended ground floors are not usually cost effective but may be used to avoid excessive fill below solid floors or on sloping ground. The under floor space should be ventilated; air bricks and sleeper walls built honeycombed with d.p.c. may be required and are measured with either floors or substructure but billed with the latter.

Timber plates and bearers are measured linear and the length measured should allow for any laps or joints required. Timber joists are measured in the same way but are described as floor members. The length taken for the joists should include for building in if required. If preservative treatment is specified to the ends of joists this is enumerated but general preservative treatment to timber would be described with the timber or as a preamble.

To ascertain the number of joists in a room, take the length of the room at right angles to the span of the joists and subtract, say, a 25 mm clearance plus half the thickness of the joist at each end. This will give the distance from the centre lines of the first and last joists which is divided by the spacing of the joists. The result, which should be rounded to the next whole number, gives the number of spaces between the joists. To this number, one must be added to give the number of joists rather than the spaces. Two points should be borne in mind before making the calculation. Firstly, the room size used for the calculation should be that of the room below the floor as the joists will be supported by the walls of that lower room. Secondly, if any of the intermediate joists are to be in a fixed position, such as a trimmer around an opening, two separate dimensions should be taken for the division on either side of the fixed joists, thus avoiding any increase in the maximum spacing. Additional joists may be required to support upper floor partitions and care must be taken to include for these. Fig. 20 shows how to calculate the number of joists required for a timber ground floor.

The trimming of timber joists for staircases, ducts, hearths, access panels, etc. is usually taken with the floor construction. Joists used in trimming should be increased in thickness by 25 mm and an addition made to their length for jointing unless metal hangers and framing brackets are used. Displaced joists and floor coverings have to be deducted.

To prevent joists twisting, herringbone or solid strutting, which may not be shown on the drawings, should be taken. This is measured linear over the joists and there should be one row, say, every 2.4 m.

Building regulations may require ties to be provided where

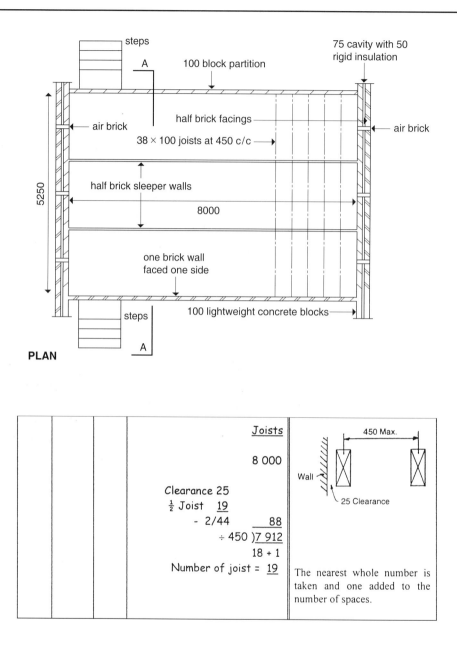

PLAN

Fig. 20

timber joists run parallel to an envelope wall; these are probably best taken with the floor construction.

Staircases

Timber staircases are enumerated and may either be described fully with dimensions or accompanied by a component detail. Items such as linings, nosings, cover moulds, trims, soffit linings, spandrel panels, ironmongery, finishes, fixings, wedges and the like, where supplied with or with part of the component, are deemed to be included. They would, of course, have to be included in the description or shown on the component detail. If these items are not part of the composite item then they are measured separately according to the appropriate rules. For example, cover fillets and nosings are measured as linear items with ends being deemed included. The creation of the stairwell is usually measured with the floors section but it may be necessary to adjust the internal finishes at this stage. Decoration on the staircase itself, if not carried out at the factory, has to be measured. Whilst there are no special rules for painting on staircases, it must be remembered that work in staircase areas has to be given separately. Handrails and balustrades that are isolated and do not form part of a staircase unit are measured as linear items and their size is stated. Ramps, wreaths, bends, and ornamental ends on handrails are enumerated as extra over the handrail.

Concrete floors

The measurement of concrete upper floors is described in Chapter 12 as part of reinforced concrete structures.

Pre-cast concrete beam and pot floors

A common ground floor construction used in UK domestic properties comprises a series of pre-cast concrete ribs (inverted T shape) supporting concrete blocks. This normally rests on block walls. The floor is measured as a composite item, including the ribs and beams as described in Fig. 21 and the sample form, below.

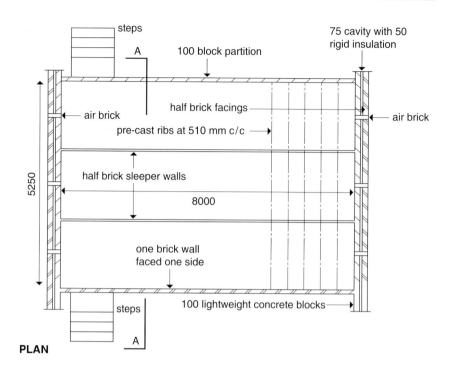

PLAN

Fig. 21

				Suspended ground floor
			Span	
			5 250	
		less supporting walls	215	
			3)5 035	
			1 676	
8.20		Precast composite slab		Floor is measured over
5.45		comprising 155mm thick		supporting walls
		overall beam and block		
		flooring, typical 510mm		
		spacing of beams, spanning		
		1.68 m, with 100mm thick		
		infill blocks (Bison or		
		similar approved)		

Example 5

Timber Upper Floor

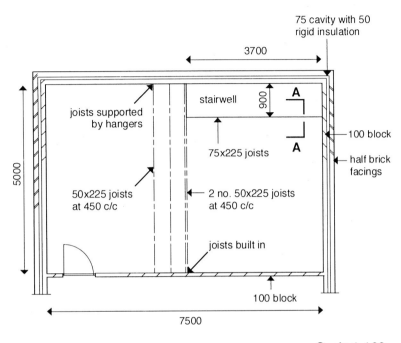

PLAN

Scale 1:100

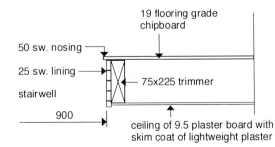

SECTION A-A

Scale 1:20

Fig. 22

Example 5

Taking-off list	SMM reference
Floor joists	G20.6.0.1
Trimmer joists	
Adjust at stairwell	
Joist hangers	G20.21.1
Herringbone strutting	G20.10.1
Steel straps	G20.20.1
Floor boarding	K20.2.1.1
Nosings	P20.2.1
Apron lining	P20.1.1
Decoration to lining	M60.1.1

				Timber Upper Floor 1

3 700

Clearance   25

Joist   ½/50 - 50

450)3 650

= 9

Where joists are in a fixed position the number of joists should be calculated from that position. In this example the stair trimming joist is used as the base.

7 500
- 3 700
3 800

As above   50

2/50   100   - 150   For trimmers

450)3 650

= 9

Length   5 000
100
5 100

²/9/ 2/	5.10 5.10		Floor members 50 x 225 mm treated sawn softwood (double trimming joist)	G20.6.0.1

900
75
975

9/	0.98		Deduct last item	Adjustment of joists at stairwell.
	3.70		Floor members 75 x 225 mm a.b.	Stair trimmer.

				Timber Upper Floor 2
9/ 2/	1 1		3 mm welded galvanised mild steel joist hangers for 50 x 225 mm members to blockwork.	G20.21.1
2/	1		Ditto for 75 x 225 mm member.	For trimmer joist.
9/	1		Ditto 50 x 225 mm but fixed to timber.	For ends of trimmed joists.
	7.50		Herringbone strutting 38 x 19 mm, treated sawn softwood to 225 mm deep joists.	G20.10.1 Taken to give side support to joists.
2/₂/	1		Galvanised mild steel joist restraining strap 30 x 5 x 1000 mm girth, once bent and 3 x countersunk drilled.	G20.20.1 These should be taken every 2 m where joists are parallel to the external wall. Fixing deemed included.

				Timber Upper Floor 3

| 7.50 | | 19 mm flooring grade | K20.2.1.1 |
| 5.00 | | chipboard exceeding 300 mm. | |

		3 700		900
		10		10
		3 710		910

3.71		Deduct
0.91		
		last item.

0.91		Nosing 50 x 25 mm wrot	P20.2.1
3.71		softwood, twice rounded	Method of jointing stated.
		tongued to edge of chipboard	Ends deemed included.
		including groove.	

| 3.70 | | Apron lining 25 x 250 mm | P20.1.1 |
| 0.90 | | wrot softwood | |

		225
	Ceiling	$9\frac{1}{2}$
	Clearance	10
		$244\frac{1}{2}$

&

		Knot, prime and stop and two	M60.1.1
		undercoat and one gloss coat	Staircase areas given
		to general surfaces n.e. 300	separately (M1).
		mm girth in staircase areas.	

Chapter 9
Roofs

Subdivision

The measurement of roofs is subdivided conveniently into two main sections on coverings and construction. As with floors, if the coverings are measured first it should be easier to understand the construction. However it is common practice to measure the construction first, starting with the wall plate and following the order in which the roof would be built. When a building has different kinds of roof, e.g. tiled roofs and felt flat roofs, each should be dealt with separately. If there are several roofs of similar type it may be more convenient to group these together. The rainwater system, unless designed with the plumbing installation, is measured with the roofs section, and may be taken either at the end of each type of roof or at the end of the entire roof measurement.

If no roof plan is provided, this may be superimposed on the plan of the upper floor. Special care must be taken to ensure that all hips and valleys are shown, one or the other being necessary at each change of direction in a pitched roof. In the case of a flat roof, it is necessary to show the direction of fall, gutters and outlets.

Pitched roof construction

The notes on timber sizes and the measurement of timber floor construction in the previous chapter should be understood before reading this section. With traditional timber roof construction a logical sequence similar to the following should be adopted:

	SMM reference
Plates	G20.8
Rafters	
Ridge	
Hips and valleys	G20.9
Ceiling joists and collars	
Purlins, ceiling beams, hangers and struts	
Sprockets	G20.17
Adjustments for openings, dormer construction, etc. (as above)	
Ties to walls	G20.20
Walking boards	K20.2
Insulation	P10.2
Ventilation	H60.10

The number of rafters in a gabled roof should be calculated by taking the total length of the roof between the centre lines of the end rafters and dividing this length by the rafter spacing. The result should be rounded to the next whole number and one added to convert the number of spaces to the number of rafters. The number of rafters will have to be multiplied by two for the two slopes. If an intermediate rafter such as a trimmer is in a fixed position then the calculation should be made on either side of the fixed rafter. If the end of the roof is hipped then a similar calculation can be made taking the dimension from the foot of the jack rafters at the hipped end. One extra rafter should be taken at each hipped end opposite the ridge as shown in Fig. 23. It will be seen that the jack rafters at the hipped end are equal to the lengths of the common rafters (shown dotted) that would have been there if the roof were gabled. The number of ceiling joists may be calculated in a similar way, but the length to be divided is taken between the inside face of the external walls less the clearance as shown for timber floor construction.

All structural timber in pitched roofs is measured linear and described as in pitched roofs, stating the size in the description. Plates and bearers are kept separately and described as plates. Cleats and sprockets are enumerated, their length and size being given in the description. Metal connectors, straps, hangers, shoes, nail plates, bolts and metal bracing are enumerated.

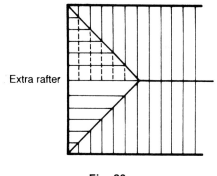

Fig. 23

Roof slopes

The roof slope of a pitched roof is the hypotenuse of a triangle, the base of which is half the roof span and the height that of the roof. In the triangle ABC in Fig. 24, AB could be found by using Pythagoras if the lengths AC and CB were known. It is more common for the span of a roof and the pitch to be given on the detail drawings.

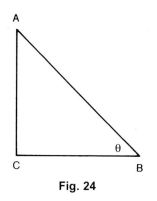

Fig. 24

Here we would use the following formula:

$$\text{cosine } \theta = \frac{BC}{AB}$$

therefore

$$AB = \frac{BC}{\text{cosine } \theta}$$

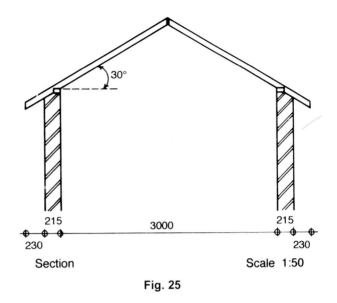

Section   Scale 1:50

Fig. 25

Consider the example in Fig. 25. Half the roof span must of course be calculated to the extreme projection of the eaves or the edge of the tiling if coverings are being measured. The span inside the walls being 3000 mm, half the span for this purpose will be 1500 + 215 + 230 = 1945 mm. The angle of pitch being 30°, the length of the slope will be:

$$1945 \div \text{cosine } 30°$$
$$= 1945 \times 1.1547$$
$$= 2246 \text{ mm}$$

and this can be checked by scaling.

It should be noted that this calculation gives the length along the top of the rafter to the centre line of the ridge. This may have to be adjusted for the length of the covering, which may project into a gutter.

Hips and valleys

The length of a hip or valley in a pitched roof must be calculated from a triangle, there usually being no true section through it from which it can be scaled. When the pitch of the hipped end is the same as that of the main roof then the length of the hip may be found from half the span and the length of the roof slope using Pythagoras.

For example in Fig. 26:

$$BE \text{ and } BD = \tfrac{1}{2} \text{ span}$$

$$AD \text{ and } AE = \text{length of slope}$$

$$\therefore AB = \sqrt{AD^2 + BD^2}$$

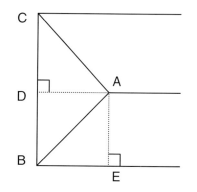

Fig. 26   Plan of hipped end to roof

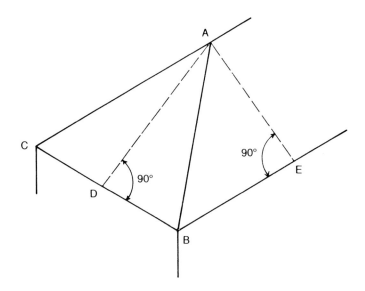

Fig. 27

Broken-up roofs

It should be noted that when a roof is broken up by hips and valleys, so long as the pitch is constant the area will always be the overall length multiplied by twice the slope. The area of tiling on a roof hipped at both ends is therefore measured in the same way as if it were gabled, the only difference being that if it were gabled the dimension of the length would probably be smaller, the projection of verges being less than that of eaves.

For example in Fig. 28:

Area of triangle ABD = area of triangle ABF

$BE = BD = AF = \frac{1}{2}$ span

AB is common

$A\hat{D}B = 90°$ and $A\hat{F}B = 90°$

With two sides and one angle equal, the triangles ABF and ABD are equal in area.

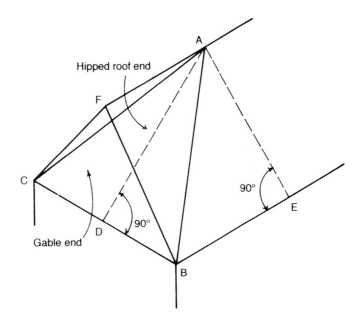

Fig. 28

Just as the length of slope is the length on plan divided by the cosine of the angle of pitch, so the area of a roof of constant pitch is the area on plan divided by the cosine of the angle of pitch; this formula can be a useful check on the measurements when worked out.

Trussed rafters

Trussed rafters are enumerated and described and although there is no specific requirement in the SMM for a dimensioned diagram to be given, this is often the clearest way to describe a truss. Bracing and binders to the construction are taken as linear items. When calculating the number of trussed rafters, consideration should be given to the position of water storage tanks and the like, as their support may require additional trusses, platforms and bracing. This will probably involve discussion with the surveyor measuring the plumbing installation; agreement should be reached on where additional items will be measured and by whom.

Tile or slate roof coverings

The first measurement in this type of roof should be the area of coverings; the description will include battens and underlay. If the area of one slope is entered in the dimensions, care must be taken to multiply this by two to allow for both slopes. Following the coverings, one should proceed around the edges of the roof, measuring abutments, eaves, verges, ridges, hips and valleys. The adjustments for chimneys and dormers and any appropriate metal gutters, soakers and flashings could conclude the coverings measurement or be taken as a separate section at the end after the construction.

Unless otherwise specified, customary girths for leadwork to pitched roofs are as follows:

- Cover flashings   150 mm
- Stepped flashings (over soakers)   200 mm
- Ditto (without soakers)   300–350 mm
- Aprons   300 mm
- Soakers 200 mm wide   length = gauge
                                           + lap + 25 mm
                                           (number = length of slope
                                           ÷ gauge).

Eaves and verge finish

Verge, eaves, fascia and barge boards are measured linear giving their size in the description. Those exceeding 300 mm wide may be measured superficial. Painting the boards should be taken at the same time as a superficial item unless the member is isolated and less than 300 mm wide when it is measured linear.

Rainwater installation

Rainwater gutters are measured linear over all fittings, these being enumerated and described as extra over the gutter. Rainwater pipes are measured in a similar manner and, in addition, stating whether they are fixed in ducts, chases, screeds or concrete. The description of gutters and rainwater pipes should include their size, method of jointing, type of fixing and whether to special backgrounds. Rainwater heads are enumerated and gratings may be included in the description.

A careful check should be made to ensure that all roof slopes are drained, that there are sufficient rainwater pipes and that the drainage plan shows provision for each pipe. If the sizes of gutters and pipes are not shown then these have to be calculated, but it should be remembered that a variety of sizes, although theoretically correct, does not always lead to an economic solution.

Flat roofs

The same main subdivision of coverings and construction can be made as before. The examples show the measurement of asphalt flat roofs. The measurement of metal covered roofs should not prove too difficult provided that their construction is fully understood.

Metal roofs are measured their net visible area plus additions for drips, welts, rolls, seams, laps and upstands as detailed in the SMM. No deductions are made for voids not exceeding $1\,m^2$ within the area of the roof. When the opening has not been deducted, no further work around the opening has to be measured except in the case of holes. The pitch of the roof has to be stated in the description.

Flashings are measured linear, stating the girth; dressing into grooves, ends, angles and intersections is deemed included. Gutters are measured in a similar manner and labours are deemed to be included. Outlets are enumerated, stating their size, and again labours are deemed included.

Asphalt roofing is measured the area in contact with the base and again no deductions are made for voids within the area which do not exceed $1\,m^2$. The pitch has to be stated in the description. The width of work has to be classified as not exceeding 150 mm, 150 to 225 mm, 225 to 300 mm, and exceeding 300 mm. Skirtings, fascias, aprons, channel and gutter linings, and coverings to kerbs are measured linear and classified in the same widths or girths, the exact girth being given if over 300 mm. Most labours to these items are deemed included; however, internal angle fillets, fair edges, rounded edges, drips and arrises (angles) on main areas are measured as linear items.

Felt roof coverings are measured superficial, stating the pitch. Skirtings, gutters, coverings to kerbs, etc. are measured linear when not exceeding 2 m girth; the girth is given in 200 mm stages. Linings, outlets and dishing to gullies are enumerated.

Timber flat roof construction is measured in a similar manner to that of floors. The fall in the roof may be created by the use of firrings, which are measured linear, stating their thickness and mean depth. Gutter boards and sides are measured superficial, classifying the width as exceeding or not exceeding 300 mm. Timber drips and rolls, if required for joints in metal coverings, are measured linear, stating their size. Fascias, soffits and rainwater goods are taken in a similar manner to those for pitched roofs.

Concrete roofs are measured in a similar manner to floors as described in Chapter 12. The fall in the roof is usually created in the screed; however, if the slab is sloping then this is classified as either exceeding or not exceeding 15°. If the slab exceeds 15°, formwork has to be measured to the upper surface. Concrete upstands are measured as cubic items; formwork to the sides, if under 1 m, is measured linear and classified as not exceeding 250 mm, 250 to 500 mm, and 500 mm to 1 m. Trowelling the surface of the concrete is measured as a superficial item.

Example 6

Traditional Pitched Roof

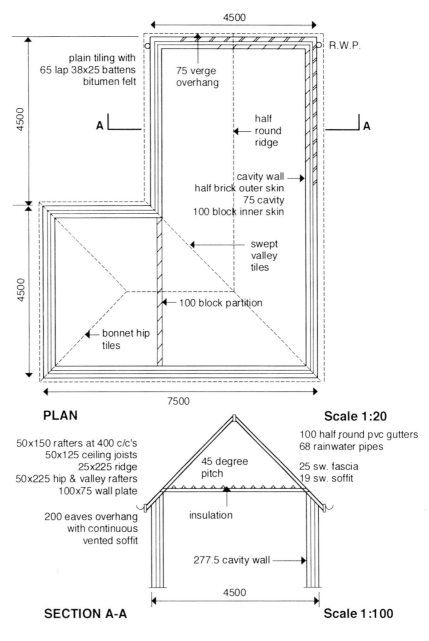

PLAN   Scale 1:20

50x150 rafters at 400 c/c's
50x125 ceiling joists
25x225 ridge
50x225 hip & valley rafters
100x75 wall plate

200 eaves overhang
with continuous
vented soffit

100 half round pvc gutters
68 rainwater pipes

25 sw. fascia
19 sw. soffit

45 degree
pitch

insulation

277.5 cavity wall

4500

SECTION A-A   Scale 1:100

Fig. 29

Example 6

Taking-off list	SMM reference
Structure	
Wall plates	G20.8.1
Roof members: rafters	G20.9.2.1
ridge, hip, valley	
ceiling joists	
Coverings	
Tiling, battens and underlay	H60.1.1
Ridge	H60.6
Saddle	H71.25.1
Hip tiles	H60.7
Valley tiles	H60.9
Verge tiles	H60.5
Eaves course	H60.4
Tilting fillet	G20.13.0.1
Fascia	G20.15.13.1
Soffit board	G20.16.3.1
Supports	G20.13.0.1
Spandril panels	
Decoration to eaves	M60.1.0.2.4
	M60.1.0.1
Eaves ventilator	H60.10
Insulation	P10.2.3.1
Rainwater goods	
Gutters	R10.10.1
Fittings	R10.11.2.1
Down pipes	R10.1.1.1
Fittings	R10.2.4.5/6
	R10.6.4.1
Joint to drain	R10.2.2.1

					Traditional Roof 1

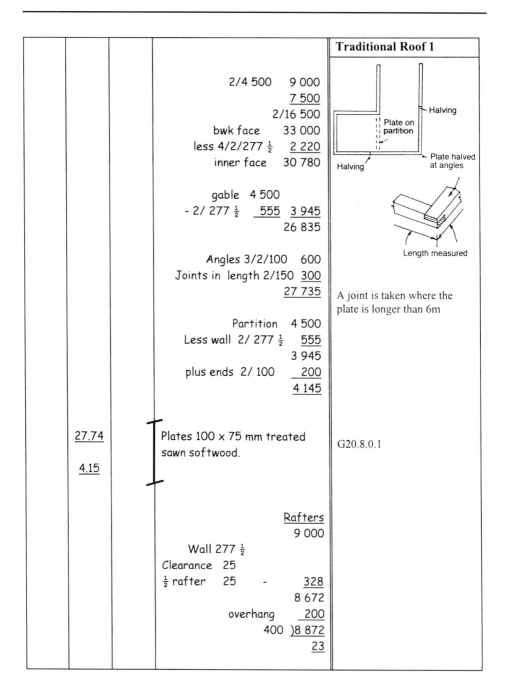

	2/4 500	9 000	
		7 500	
	2/16 500		
	bwk face	33 000	
	less 4/2/277 ½	2 220	
	inner face	30 780	
	gable	4 500	
	- 2/ 277 ½	555	3 945
			26 835
	Angles 3/2/100	600	
	Joints in length 2/150	300	
		27 735	
	Partition	4 500	
	Less wall 2/ 277 ½	555	
		3 945	
	plus ends 2/ 100	200	
		4 145	

A joint is taken where the plate is longer than 6m

27.74	Plates 100 x 75 mm treated sawn softwood.	G20.8.0.1
4.15		

	Rafters	
		9 000
Wall 277 ½		
Clearance 25		
½ rafter 25	-	328
		8 672
overhang		200
400	)8 872	
		23

				Traditional Roof 2

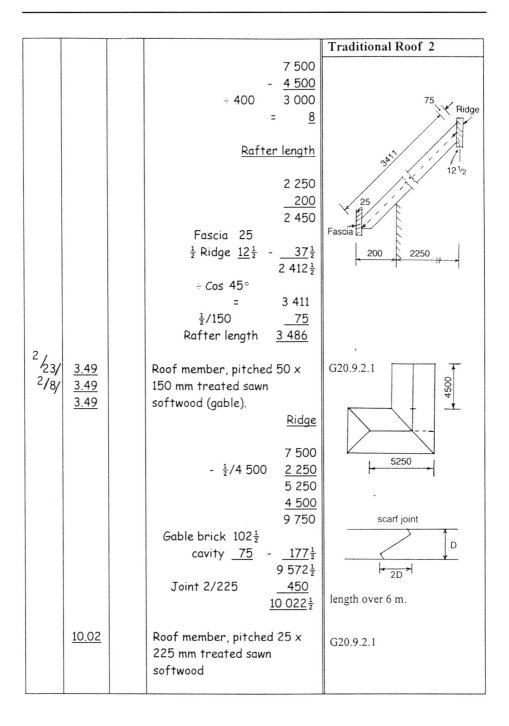

$$7\ 500$$
$$-\ \underline{4\ 500}$$
$$\div\ 400 \qquad 3\ 000$$
$$=\qquad \underline{8}$$

Rafter length

$$2\ 250$$
$$\underline{200}$$
$$2\ 450$$

Fascia 25
½ Ridge 12½ - $\underline{37\tfrac{1}{2}}$
$$2\ 412\tfrac{1}{2}$$

÷ Cos 45°
$$=\qquad 3\ 411$$
½/150 $\qquad \underline{75}$
Rafter length $\underline{3\ 486}$

| 2/23/ 2/8/ | 3.49 3.49 3.49 | | Roof member, pitched 50 × 150 mm treated sawn softwood (gable). | G20.9.2.1 |

Ridge

$$7\ 500$$
$$-\ \tfrac{1}{2}/4\ 500 \quad \underline{2\ 250}$$
$$5\ 250$$
$$\underline{4\ 500}$$
$$9\ 750$$

Gable brick 102½
cavity $\underline{75}$ - $\underline{177\tfrac{1}{2}}$
$$9\ 572\tfrac{1}{2}$$
Joint 2/225 $\qquad \underline{450}$
$$10\ 022\tfrac{1}{2}$$

length over 6 m.

| | 10.02 | | Roof member, pitched 25 × 225 mm treated sawn softwood | G20.9.2.1 |

				Traditional Roof 3
			$3\,486^2 + 2\,412^2 = \text{Hip}^2$	

$$\text{Hip} = \underline{4\,239}$$

$\dfrac{3}{1}/$ 4.24 Roof member pitched 50 x 225 mm treated sawn softwood.

<u>Ceiling Joists</u>

```
                        9 000
      Wall   277½
  Clearance      25
  ½ Joist ½/50 25
        - 2/327½ -   655
                   8 345
    ÷ 400 =        21 + 1
          =           22

                    7 500
                  - 4 500
                    3 000
                  -  327½
                    2 672½

      102½
       75            177½
                    2 850

       25
   ½/50 25         -   50
                    2 800

     ÷ 400 =        7 + 1
         =            8
```

				Traditional Roof 4

30/	4.50		Roof member, pitched, 50 x 125 mm treated sawn softwood

G20.9.2.1

Ceiling joists   22

           +   8

               30

```
                        4 500
         - 2/ 327 ½      655
                  400 )3 845
                    10 + 1
                        11
```

Where joists rest on partition at junction of L shape

11/	0.15		<u>Deduct</u> last item

<u>Coverings</u>

```
                    7 500
Overhang  200
Into gutter   50
    2/ 250          500
                  8 000
   plus return    4 500
   length =      12 500
```

Rafter

45°

Clg. Jst.

Plate

200

<u>slope length</u>

```
                  4 500
eaves a.b.         500
            2 ) 5 000
½ span =    2 500

slope =     2 500
             cos θ

       =    3.535
```

				Traditional Roof 5
2/	12.50 3.54		Roof coverings, 45 ° pitch , handmade sand faced plain clay tiles size 265 x 165 mm, 65 mm end lap, nailed every fourth course with two aluminium nails, 38 x 25 mm treated softwood battens with galvanised nails and reinforced bitumen felt to BS747 type 1F weighing 15 kg/m² with 75 mm horizontal and 150 mm end laps	H60.1.1 The area of the roof coverings does not need to be adjusted for the hipped end, as long as the pitch on the hipped end is the same as the main slope.

$$
\begin{array}{rr}
 & \underline{\text{Ridge}} \\
\text{2/4 500} & \text{9 000} \\
\text{Eaves 250} & \\
\text{verge} \quad \underline{75} & \underline{325} \\
 & \text{9 325} \\
+\,\text{return} & \underline{8\ 000} \\
 & \text{17 325} \\
-\,3/\,\tfrac{1}{2}\,/5\ 000 & \underline{7\ 500} \\
 & \text{9 825}
\end{array}
$$

	9.83		Ridge 250 mm diameter half round, red clayware	H60.6
2/	1		Saddle 450 x 450 mm Code 4 lead, supply only & Fix only ditto	H71.25.1 At angle of ridge and junction with hips H60.10.6.1.1

				Traditional Roof 6
			Hips	
			Using pythagoras the hip	
			length is $\sqrt{(3.535^2 + 2\,500^2)}$	
			= 4 330	
3/	4.33		Hip matching bonnet hip tiles	H60.7
	4.33		Purpose made swept valley tiles.	H60.9
2/	3.54		Verge with plain tile underclaok	H60.5 C2 states undercloaks deemed included but S3 requires method of forming to be stated.
			Eaves	
			8 000	
			9 325	
			2/17 325	
			34 650	
			less gable 4 500	
			eaves 2/250 500 5 000	
			29650	
	29.65		Eaves double tile course	H60.4
			Tilting fillet	
			29 650	
			less overhang 2/2/ 50 200	
			29 450	
	29.45		Individual support 75 x 50 mm extreme) treated softwood triangular	G20.13.0.1

Diagram labels: 3535, 3535, 2500, 2500, Hip

				Traditional Roof 7
	29.45		Fascia board 25 x 175 mm wrought softwood, chamfered and grooved	G20.15.3.1

Soffit
Fascia length 29 450
+ for int. corners 2/200 400
29 850

Face of wall

Soffit board

| | 29.85 | | Eaves soffit board, 19 x 200 mm wrought softwood, rebated | G20.16.3.1 |

bwk face   33 000
less gable   4 500
400) 28 500
72 + 1
= 73

Brackets at rafter position for soffit fixing

73/	0.20		Individual support 38 x 50 mm treated sawn softwood	G20.13.0.1 Horizontal
73/	0.18			vertical
2/	1		Spandril boxed end to eaves 25 mm wrought softwood size 200 x 270 mm overall	Filling projecting eaves at gable end
	29.45		Prime only wood general surfaces n.e. 300 mm girth before fixing	
	29.85			
2/	1		Ditto isolated areas n.e. 0.50m^2	

				Traditional Roof 8
	29.85 0.40		Knot, prime and stop and two undercoats and one gloss coat to wood general surfaces exceeding 300 mm externally. 33 000 - 4 500 28 500	M60.1.0.1 Fascia and soffits
	28.50		Proprietary continuous eaves ventilator fixed to softwood. _Gutter_ Fascia 29 450 3/2/100 600 30 050	H60.10
	7.35 3.95 2.60 3.95		Insulation 100 mm thick laid between joists at 400 mm centres, horizontally.	R10.2.3.1
	30.05		Rainwater gutters, straight half round 100 mm UPVC with combined fascia brackets and clips screwed to softwood.	R10.10.1.1 Measured over fittings
2/	1		Extra over ditto for stop end.	R10.11.2.1
4/	1		Ditto angles.	
4/	1		Ditto outlets.	

				Traditional Roof 9
4/	6.00		Rainwater pipes, straight 68 mm diameter UPVC with push fit socket joints and ear piece brackets at 2 m centres plugged to masonry.	R10.1.1.1 Pipe length taken over fittings. Length assumed
4/	1		Extra over 68 mm UPVC for shoe with push fit socket joint.	R10.2.4.6 Fittings over 65 mm diameter have to be described.
4/2/	1		Ditto for offset bend with ditto.	R10.2.4.5 Bends to form swan-neck.
4/	1		Copper wire balloon grating to 68 mm diameter pipe. & Joint to top inlet of clayware gully with cement and sand (1:3).	R10.6.4.1 R10.2.2.1

Example 7

Trussed Rafter Roof

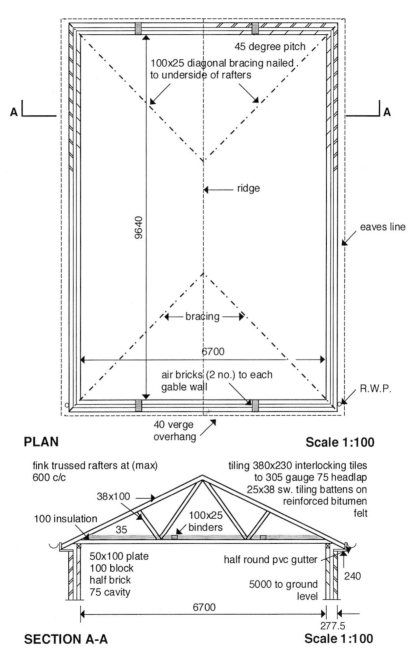

PLAN Scale 1:100

SECTION A-A Scale 1:100

Fig. 30

Example 7

Taking-off list	**SMM reference**
Structure	
Wall plates	G20.8.1
Trusses	G20.2.1
Binders	G20.9.2.1
Bracing	
Joist anchors	G20.20.1
Insulation	P10.2.3.1
Air bricks	F30.12.1
Coverings	
Tilings, battens and underlay	H60.1.1
Ridge	H60.6
Verge tiles	H60.5
Eaves course	H60.4
Tilting fillet	G20.13.0.1
Fascia	G20.15.13.1
Soffit board	G20.16.3.1
Supports	G20.13.0.1
Spandril panels	
Decoration to eaves	M60.1.0.2.4
	M60.1.0.1
Eaves ventilator	H60.10
Rainwater goods	
Gutters	R10.10.1
Fittings	R10.11.2.1
Down pipes	R10.1.1.1
Fittings	R10.2.4.5/6
	R10.6.4.1
Joint to drain	R10.2.2.1

				<u>Plates</u>	**Trussed Roof 1**
				9 640	
			joint	<u>150</u>	
				9 790	
2/	<u>9.79</u>		Plates 100 x 75 mm treated sawn softwood.		G20.8.0.1
				<u>trusses</u>	
				9 640	
			Clearance 25		
			½ rafter <u>19</u>		
			- 2/44	<u>88</u>	
			600	)9 552	
				16 +1	
				=<u>17</u>	
				<u>Overhang</u>	
			Eaves	240	
			Wall	<u>277</u>	
				517	
			less fascia	<u>25</u>	
				<u>492</u>	
17/	<u>1</u>		Trussed rafters in sawn softwood FINK pattern comprising 38 x 100 mm members. 6700mm clear span, 492 mm overhang, 35 ° pitch, jointed with 18 g galvanized steel plate connectors.		G20.2.1 A dimensioned diagram could be included for this item.

			<div align="right">Bracing</div>	**Trussed Roof 2**

½ span 6.700   3 350
plate   100
length on plan   3 450

Roof slope = 3 450 / cos θ
= 4 212

length =√ (3.45² + 4.212²)
= 5 445

3/	9.64		Roof members pitched 25 x 100mm treated sawn softwood.	G20.9.2.1 Binders
2/2/	5.45			Bracing
2/2/	1		Galvanised mild steel joist anchor restraint 30 x 5 mm x 300 mm girth, twice bent and twice countersunk drilled.	G20.20.1

9 640
less trusses   17/38   646
8 994

	8.99		100 mm insulation between members at 600mm centres horizontally.	P10.2.3.1
	6.70			
2/2/	1		Airbricks, square hole pattern, red terracotta size 225 x 150mm in hollow wall of ½ b facings, 100mm blks and 75 mm insulated cavity including sealing cavity with slates	

Roof members pitched 25 x 100mm treated sawn softwood.

Bracing
4212
Brace
3450

				Trussed Roof 3

<u>Covering</u>

		9 640
Walls 2/277½		555
		10 195
Verge 2/40		80
		10 275

Wall 277½	6 700
Eaves 240	
Into gutter 50	
2 /567½	1 135
	½/7 835

= 3918 ÷ cos 35°

Slope length = <u>4 783</u>

The slope length of the roof equals half the span multiplied by the secant of the pitch angle.

2/	10.28		Roof coverings, 35° pitch, 380 x 230 mm Redland Interlocking tiles to 75 mm lap, nailed every second course with 48 x 11 gauge aluminium alloy nails, 38 x 25 mm battens galvanised nailed, reinforced bitumen roofing felt to BS747 type 1F weighing 15 kg with 75 mm horizontal and 150 mm end laps.	H60.1.1
	4.78			

				Trussed Roof 4
2/	10.28		Eaves tiling & Individual support 75 x 50 mm extreme, treated softwood, triangular	H60.4 G20.13.0.1
2/2/	4.78		<u>Verge</u> Verge, special tile including plain tile undercloak.	H60.5
	10.28		<u>Ridge</u> Ridge, 250mm diameter half round to match tiling.	H60.6
2/	10.20		Fascia board 25 x 150mm wrot softwood, chamfered and grooved.	G20.15.3.1

25	240
- 13 -	12
	228

&

Eaves soffit board 25 x 228 mm wrot softwood, rebated.   G20.16.3.1

				Trussed Roof 5
2/	<u>10.20</u>		Individual support 38 x 50 mm treated sawn softwood to brick.	G20.13.0.1
2/2/		<u>1</u>	Spandril boxed end to eaves 25mm wrot softwood size 240 x 300 mm overall.	
2/2/	<u>10.20</u>		Prime only general surfaces of wood n.e. 300 mm wide, before fixing.	M60.1.0.2.4 Priming backs of fascia and soffit before fixing
2/2/		<u>1</u>	Ditto isolated areas n.e. 0.5 m².	M60.1.0.3.4 Spandril ends

$$\begin{array}{r} 10\ 195 \\ \text{Ends } 2/300 \quad \underline{600} \\ \underline{10\ 795} \end{array}$$

$$\begin{array}{r} 250 \\ 150 \\ \underline{25} \\ \underline{425} \end{array}$$

2/	10.80 <u>0.43</u>		Knot, prime, stop, two undercoats and one gloss coat to wood general surfaces exceeding 300 mm, externally.	M60.1.0.1 Taken as superficial work to both surfaces as same treatment.

				Trussed Roof 6
2/	10.28		Rainwater gutters, Straight 100 mm half round PVCu with combined fascia brackets and clips, screwed to softwood.	R10.10.1.1 Measured over fittings
2/2/	1		Extra over for stop end.	R10.11.2.1
2/	1		Ditto for outlet. & Copper wire balloon grating to 68 mm diameter outlet. & Joint to top inlet of clayware gully with cement and sand (1:3)	 R10.6.4.1 R10.2.2.1
2/	5.20		Rainwater pipes, straight 68 mm diameter PVCu with push fit socket joints and earpiece brackets at 2 m centres plugged to brickwork.	R10.1.1.1 Pipe length assumed and measured over fittings
2/2/	1		Extra over for 68 mm diameter r.w.p. for offset bend.	R10.2.4.5 Bends to form swan neck. Fittings to pipes over 65 mm diameter must be described.

Example 8

Asphalt Covered Flat Roof

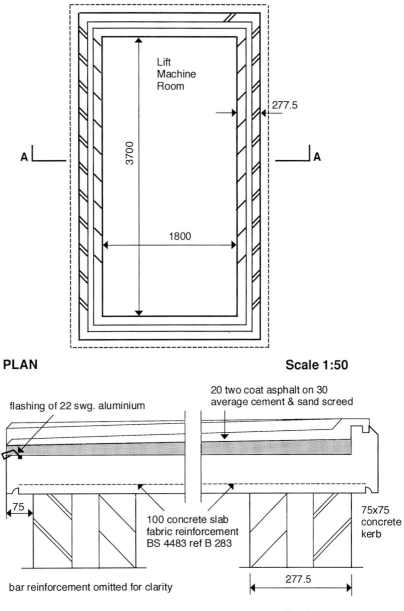

PLAN Scale 1:50

flashing of 22 swg. aluminium

20 two coat asphalt on 30 average cement & sand screed

75

100 concrete slab fabric reinforcement BS 4483 ref B 283

75x75 concrete kerb

277.5

bar reinforcement omitted for clarity

SECTION A-A Scale 1:10

Fig. 31

Example 8

Taking-off list	SMM reference
Concrete slab	E10.5.1.0.1
Concrete upstand	E10.14
Soffit formwork	E20.8.1.1
Fair finish	E20.20.1
Groove	E20.17.1.0.1
Edge formwork	E20.3.1.2
Upstand formwork	E20.4.1.2
Recess	E20.17
Surface treatment to upstand	E41.3
Reinforcement	E30.4.1
Asphalt roofing	J21.3.4
Concrete surface treatment	E41.3
Screed	M10.6.1
Asphalt upstand	J21.11.1
Fair edge	J21.13/14
Flashing	H72.10.1.0.1
Fixing bearer	G20.13.0.1

				Size of slab	Flat roof 1
				1 800 3 700	
			2/277 ½	555 555	
			2/ 75	150 150	
				2 505 4 405	
2.51 4.41 0.10			In situ concrete (25 N – 20 mm aggregate) slab n.e. 150 mm thick, reinforced.		E10.5.1.0.1
				4 405	Upstand goes around two short sides and one long side.
			2/2 505	5 010	
				9 415	
			less 2/ 75	150	
			centre of upstand	9 265	
			Ditto, upstand		
9.27 0.08 0.08					E10.14
				Formwork	
				2 505	
				4 405	
				2/ 6 910	
			external girth =	13 820	
			- 4/2/ ½ /75 =	300	
				13 520	
3.70 1.80			Formwork, soffits of slabs n.e. 200 mm thick, horizontal, height 9 –10.5 m.		E20.8.1.1 Support height is given in stages of 1.5 m. It is assumed that this slab is above a lift shaft.
13.52 0.08			&		
			Extra over formwork to soffit of slabs for fair finish		E20.20.1

					Flat Roof 2
	13.52		Ditto for throating 25 mm radius		E20.17.1.0.1
	13.82		Formwork to edge of suspended slab, vertical n.e. 250 mm high		E20.3.1.2 Taken to outer edge of upstand total height not less than 250 mm.
	4.51 0.10		Extra over formwork to edges of slab for fair finish		E20.20.1
	4.51 0.18				
2/	2.51 0.18				

Upstands

External face a.b.   9 415
- 2/2/75   300
inner face   9 115

				Flat Roof 3
9.12			Formwork, sides of upstands, plain vertical n.e. 250 mm high.	E20.4.1.2 Inner face of upstand
9.27			Formwork, recess, plain rectangular 20 x 20 mm in top of upstand (No 3)	Similar to E20.17 Not extra over as no formwork measured to top of upstand.
9.12 0.03			Trowelling top of upstand to form weathering	E41.3 Could be measured under E20.11 as top formwork over 15° Note:- It may be better to take all the items of formwork to the roof projection and upstand in one item as a complex shape under E20.28 and give a dimensioned diagram.

					Flat Roof 4
				<u>Reinforcement</u>	
				2 505 4 405 - Cover <u>70</u> - 2/35 <u>70</u> 2 435 4 335	
2.44 <u>4.34</u>				Reinforcement for in situ concrete, welded fabric, B283 weighing 3.73 kg/m² with 200 mm laps.	E30.4.1
				<u>Coverings</u>	
				<u>Mastic asphalt limestone aggregate to BS6925 in two coats laid breaking joints and 20 mm thick</u>	A heading can be used to cover the common parts of descriptions which would apply to several items.
				2 505 4 405 - <u>75</u> - 2/75 <u>150</u> 2 430 4 255	
2.43 <u>4.26</u>				Asphalt roofing exceeding 300 mm wide to level and falls to screeded base & Trowel surface of concrete for screed	J21.3.4 E41.3

				Flat Roof 5
2.43 4.26			Cement and sand (1:4) screed, roofs, level and to falls n.e. 15°, 30 mm average thickness in one coat to concrete.	M10.6.1
9.12			Asphalt to kerb n.e. 150 mm, raking.	J21.11.1 Deemed to include arrises, angle fillets, turning into grooves, ends and angles.

$$
\begin{array}{r}
3\ 700 \\
2/277\tfrac{1}{2}\quad 555 \\
\hline
4\ 255
\end{array}
$$

4.26			Fair rounded edge to asphalt & 22 SWG flashing in aluminium n.e. 100 mm girth, horizontal fixed with galvanised nails to hardwood. & Treated hardwood individual support 40 x 25 mm dovetail section and setting in unset concrete.	J21.13/14 (C2a) Deemed to include working to metal. H72.10.1.0.1 G20.13.0.1

Chapter 10
Internal Finishes

Schedules

A schedule of internal finishes listing floor by floor and room by room the ceiling, wall and floor finishes and decorations is an invaluable aid to the measurement of this section of the work. Details of cornices, skirtings, dados, etc. can be added to the schedule as required. The schedule brings to light any missing information, enables rooms with similar finishes to be grouped together more easily and reduces the need to refer to the specification during the taking off. Looping through the entries on the schedule as the items are measured ensures that none are missed. At a later stage the schedule provides a useful quick reference indicating what has been taken without one having to look through the dimensions. An example of a typical schedule is given in Fig. 32. This is an information schedule and may well be provided by the architect. Another schedule could be utilised to measure the finishes. Here, girths and heights of rooms would be recorded and summarised to reduce any repetitive dimensions.

Subdivision

If the ceiling heights and construction vary throughout the building, it may be advisable to keep the measurements on each floor separate. If, however, the room layout and finishes repeat from floor to floor, it is probably more expedient to use timesing. As this will quite possibly be the largest section of taking-off, care should be taken to signpost the dimensions to enable measurements to be traced easily at a later stage. The principal subdivisions in the order recommended for measuring are:

- Floor finishes and screeds (unless taken with floor construction)
- Ceiling finishes (including attached beams with the same finish)
- Isolated beams
- Wall finishes (including attached columns with the same finish)
- Isolated columns
- Cornices
- Dados
- Skirtings.

Note: It may be convenient to measure floors with ceilings if the areas are the same. Confusion may arise, however, if the specification changes are not consistent between the two and also if deductions have to be made from floors only for such items as hearths, stair openings, and fittings.

INTERNAL FINISHES SCHEDULE					
	CEILING		WALLS		
LOCATION	FINISH	DECORATION	FINISH	DECORATION	DADO
Bathroom Kitchen etc.	9.5 mm plasterboard 5 mm skim	2 ct emulsion	13 mm two ct plaster	2 ct emulsion	Glazed tiling 1 m high

(*Internal finishes schedule contd.*)					
			FLOOR		
LOCATION	CORNICE	SKIRTING	BED	FINISH	REMARKS
Bathroom Kitchen etc.	250 mm gth moulded	25 × 100 mm rounded s/w Kps & 3	35 mm cement and sand	Carpet	Area ne 4 m^2

Fig. 32

Generally

The measurements for in situ finishes are taken as those of the base to which the finish is applied. Tiling, on the other hand, is measured on the exposed face. In practice this makes little difference, as normally the structural room sizes are taken for the measurement of finishes and decoration. One should be careful, however, when measuring tiling to columns as the face area of the tiling will be greater than the structural face area. Generally, no deduction is made for voids within the area of the work when they do not exceed $0.5\,m^2$. Rounded internal and external angles exceeding 10 mm but not exceeding 100 mm radius to in situ finishes are measured as linear items. When the radius exceeds 100 mm the work becomes curved. In the case of tiling, these angles are measured only if special tiles are used.

Floor finishes

In situ sheet and tile finishings all have to be described as follows:

- Level or to falls less than 15°
- To falls and crossfalls less than 15°
- To slopes more than 15°

For in situ sheet and flexible tile floorings, the area measured is that in contact with the base. For rigid tile flooring the finished face is measured, although in practice, as far as floors are concerned, this usually makes little difference. No deduction is made for voids not exceeding $0.5\,m^2$ within the areas of the work. If the work is laid in bays this has to be stated, giving their average size. Work in staircase areas and plant rooms has to be so described.

Skirtings are usually measured with internal finishes, as their length relates to that of the walls. Floor coverings in door openings are taken either with doors, as the surveyor dealing with these is familiar with the dimensions, or with the finishes. It should be remembered that dividing strips may be required where the finish changes. Doors are usually hung flush with the face of the wall of the room into which they open; therefore the floor finish in the opening should be that of the room that shows when the door is closed. Mat frames and mat wells should be measured with floor finishes.

Screeds, if required, are measured with floor finishes as their areas are usually the same. Care should be taken to ascertain the thickness of screeds where the floor finishes vary. Usually it is necessary to have floor surfaces level; differing finishes may have varying thicknesses, requiring the screed to take up the difference. Sometimes this difference is allowed for in the floor construction.

Timber boarding, plywood or chipboard to floors is measured superficial unless it does not exceed 300 mm wide when it is measured linear. Areas not exceeding 1 m^2 are enumerated. No deduction is made for voids not exceeding 0.5 m^2 within the area of the flooring. Remember that nosings and margins may be required around openings in the floor.

Ceiling finishes

Ceiling areas are taken from wall to wall in each room using the figured structural dimensions. The dimensions are usually best taken over the beams and the adjustments made later. Care should be taken as the work to the soffit of beams may well be in a different height classification to the main ceiling of a room. If the majority of ceiling finishings are identical it may be advantageous to measure the ceiling over all the internal walls and partitions. From this measurement will be deducted the total lengths of the internal walls and partitions, possibly obtained from previous dimensions, squared by their respective thicknesses. Treating the surface of concrete by mechanical means, or as it is traditionally known *hacking concrete* to receive plaster, is measured superficially and is conveniently *anded on* to the finishes dimensions. The application of bonding agents or other preparatory work is described with the plasterwork. Plasterboard linings to ceilings are measured in square metres.

Wall finishes

In the case of wall finishings it is best to schedule or total on waste the perimeters of all rooms having the same height and finish. The areas of the walls may thus be contained in two or three dimensions instead of the large number there would be if each room were measured separately. As in the case of the measurement of walls, openings are generally ignored, the wall finish being measured across. Deductions for openings together with work to reveals would be taken later with the measurement of windows, doors and blank openings. If large openings or infill panels occur from floor to ceiling height it is sometimes preferable to measure the finishes net. Care must be taken to inform other measurers which openings have been measured net so as to avoid the possibility of a double deduction. As a general rule, if the walling has been measured across an opening, so should the finishes.

If in situ wall finishes are applied to brick or block walls, then the raking out of the joints to form a key is deemed included. Preparation of concrete walls is dealt with as mentioned for ceilings above.

Angle beads, etc.

Metal angle, casing and movement beads are measured as linear items. Sizes are given in the description but working finishes to beads are deemed included.

Decoration

The decoration to ceilings and walls should be measured with the finish. If the decoration is all the same this can be anded on to the plaster or plasterboard description. If, however, there is a mixture of several types of decoration, such as emulsion, gloss paint and wallpaper, it will probably be best to subdivide the plaster or plasterboard measurement so that the appropriate decoration can be anded on. Where all is, say, emulsion except wallpaper in one or two rooms, it is usually more satisfactory to measure the whole as emulsion in the first instance and then to deduct the emulsion and add wallpaper for the appropriate area. Isolated paintwork where the girth does not exceed 300 mm or where the area does not exceed $0.5 \, m^2$ is measured linear or enumerated, respectively. Paintwork to walls, ceilings, beams and columns is described as work to general surfaces and, if the same, can be added together in the bill. Work to ceilings and beams exceeding 3.5 m above floor level has, however, to be so described in stages of 1.5 m. Papering has to be described in the two categories of ceilings with beams and walls with columns, the same rule regarding high ceilings applying.

Cornices and coves

The measurement of cornices and coves is usually straightforward, as they normally run all round the room without interruption. Both in situ and prefabricated types are measured the length in contact with the base, using the length already calculated for the perimeter of the walls. The measurement must include returns to projections and to beams if these occur in the ceiling. Ends, angles and intersections are enumerated as extra over the items. Paintwork to cornices does not have to be kept separate if it is the same as that to the walls or ceilings. In fact it is questionable whether any adjustment should be made to the wall and ceiling decoration if this is the same, particularly if the cornice is contoured.

Skirtings

There is a difference of opinion as to whether skirtings should be measured net or measured gross across openings. The disadvantage with measuring net is that the taker-off measuring finishes will have to ascertain the width of openings and possibly the depth of reveals. On the other hand, wall decoration, as opposed to plaster, should be deducted behind the skirting. Thus, if the skirtings are measured gross with the deduction of decoration behind, when the doors are measured a deduction of plaster and decoration is usually made for the height of the opening. This results in the decoration being deducted twice behind the skirting and therefore an addition of decoration will have to be made for the length by the height of the skirting across the door opening.

Timber skirtings are measured linear, usually using structural wall face dimensions, but care will have to be exercised with measurements to isolated piers etc., as the length will be increased. The size and shape of the skirting and method of fixing if specified are given in the description but mitres, ends and intersections are deemed included except when in hardwood of more than $0.003\,\text{m}^2$ sectional area. If the skirting is fixed to grounds, no deduction of plaster is made for grounds. Dado and picture rails are measured in the same way as skirtings.

In situ and tile skirtings are measured as linear items, stating the height and thickness in the description. Fair and rounded edges are deemed to be included, as are ends and angles.

Wall tiling

Firstly, it is necessary to ascertain whether the wall tiling is fixed with adhesive or bedded in mortar. The specification may, particularly in the case of a tiled dado, call for the wall behind the tiling to be plastered like the remainder of the wall and for the tiles to be fixed to the plaster with adhesive. Alternatively, there may be a cement and sand screed behind the tiling, with the tiles either bedded in mortar or fixed with adhesive. If there is a mixture of wall tiling and plaster finish on a wall, it is usually necessary for the back of the tiles to be flush with the face of the plaster. If special tiles with rounded edges are specified at top edges and external angles these are measured as linear items extra over the main work. The rules for measurement of tiling follow closely those for in situ finishes but attention is drawn to the paragraph on p. 163 headed 'Generally' relating to measurements.

Staircase areas

Generally, work in staircase areas has to be kept separate and the rules relating to the height of the work do not apply. When staircases are contained within lobby areas, there is little difficulty in deciding what is and what is not a staircase area. When they are not, it is not always clear and in such cases the surveyor has to use his or her discretion and decide the subdivision. This decision should be passed on to the estimator by way of the bill so that later there is no argument.

Internal partitions

The measurement of timber stud partitions is described in Chapter 7. Metal stud partitions can be measured in a similar way.

Dry wall linings

These are considered to be beyond the scope of this book, but, briefly, open spaced grounds over 300 mm wide are measured superficial, stating their size, spacing and fixing. Plasterboard dry linings are measured linear, stating the height in stages of 300 mm in the description. Internal and external angles are measured as linear items.

Example 9

Internal Finishes

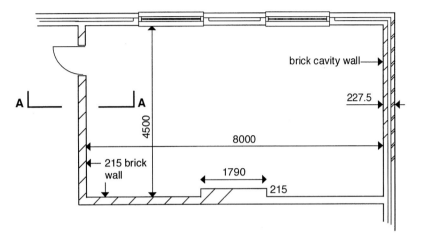

PLAN   **Scale 1:100**

brick cavity wall

227.5

4500

8000

215 brick wall

1790

215

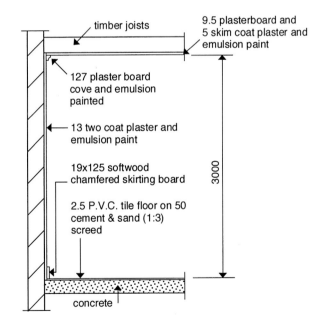

timber joists

9.5 plasterboard and 5 skim coat plaster and emulsion paint

127 plaster board cove and emulsion painted

13 two coat plaster and emulsion paint

19x125 softwood chamfered skirting board

2.5 P.V.C. tile floor on 50 cement & sand (1:3) screed

3000

concrete

SECTION A-A   **Scale 1:50**

Fig. 33

Example 9

Taking-off list	SMM reference
Ceilings: plasterboard and plaster	M20.2.1.2
emulsion	M60.1.0.1
Floor: floor tiles	M50.5.1.1
screed	M10.5.1
Wall plaster	M20.1.1.1
Plaster to pier	M20.1.2.1
Emulsion	M60.1.1.1
Angle beads	M20.24
Plaster cove	M20.17
Angles	M20.23.2/3
Skirting	P20.1.1
Paint to skirting	M60.1.0.2
Adjust paint for skirting	

				Internal Finishes 1
8.00 4.50			Plastered ceilings, width exceeding 300 mm. 9.5 mm gypsum plasterboard fixed with rust-proofed nails to softwood, 5 mm one coat thistle board finish plaster.	M20.2.1.2 Reinforcing scrim to plasterboard deemed included M20.C3. Note that it may be necessary to measure noggins between the timber joists for fixing the edge of the plasterboard.
			&	
			Two coats emulsion paint general plaster surfaces, girth exceeding 300 mm.	M60.1.0.1
			&	
			PVC tiles 225 x 225 x 2.5 mm, width exceeding 300 mm, level, bedded with bitumen adhesive to cement screed.	M50.5.1.1 The floor is 'anded-on' to the ceiling in this case as the areas are identical.
			&	
			Cement and sand (1:4) screed to floors, level, 50 mm thick in one coat to concrete base.	M10.5.1 An item of 'trowelling surface of unset concrete to receive paving' should be taken if not measured with the slabs. (E41.3).
1,79 0.22			Deduct [Pier]	Deduction for area displaced by pier. Note that the minimum deduction rule does not apply as the void is at the perimeter. GR3.4

				Internal Finishes 2

$$8\ 000$$
$$\underline{4\ 500}$$
$$2/12\ 500\ =\qquad 25\ 000$$
$$2/215\qquad\qquad \underline{\quad 430}$$
$$25\ 430$$

25.00 3.00	Plastered walls exceeding 300 mm wide in two coats 13 mm thick to brick.	M20.1.1.1 Raking out joints to form key is deemed included. F10 C1(d)
2/ 3.00	Ditto n.e. 300 mm [Pier]	M20.1.2.1 To sides of projection.
25.43 3.00	Two coats emulsion paint a.b.	M60.1.1.1 As narrow width is not isolated then area is included with main area.
2/ 3.00	Accessories, perforated metal angle bead vertical, with 50 mm returns, to brickwork with masonry nails.	M20.24 Labour on plaster to beads is deemed included. Note:- Plaster angles are deemed included (C4)

				Internal Finishes 3
			Coves	
	25.43		Gyproc cove 127mm girth fixed with adhesive.	M20.17 No adjustment for emulsion paint as this is regarded as equal to that displaced.
6/	1		Extra over ditto for internal angles.	M20.23.2
2/	1		Ditto for external angles.	M23.3
			Skirting	
	25.43		Wrot softwood skirting 19 x 125 mm chamfered plugged and screwed to brickwork.	P20.1.1 See P20 C1 for mitres. Describe method of fixing, if not at Contractor's discretion.
			&	
			Knot, prime, stop and three coats gloss paint on general surfaces woodwork, isolated surfaces n.e. 300 mm girth.	M60.1.0.2

				Internal Finishes 4
25.43			Prime only general surfaces of woodwork, n.e. 300 mm wide, prior to fixing & <u>Deduct</u> Two coat emulsion paint to plaster surfaces x 0.13 = _____ <u>Note</u>:- To take adjustment for openings.	M60.1.0.2.4 Priming back of skirtings before fixing. Normally adjustment of finishes for door and window openings is taken with these items.

Chapter 11
Windows and Doors

Subdivision

The measurement of windows and doors can be conveniently subdivided as follows:

- Windows
- External doors
- Internal doors
- Blank openings.

Whilst it is not essential to take external and internal doors separately, it will probably be found that they are different types of doors with different finishes and therefore they may have little in common.

Windows
- Timber, metal or UPVC casement and fixing
- Glass (if not included in the above)
- Ironmongery (ditto)
- Decoration (if required)

Opening adjustments
- Deduction of brickwork and blockwork and external and internal finishes
- Support to the work above the window
- Damp proofing and finishes to the head externally and internally
- Ditto to the external and internal reveals
- Ditto to the external and internal cills.

External doors
- The same items would be measured for external doors as shown above for windows.

Internal doors and blank openings
- Again this will be similar as given for Windows but in place of cills, flooring in the opening will be required. Damp proofing around the opening will not be necessary.

Steps to external doors may be measured at this stage along with porches, canopies or other special features.

If these items are followed through systematically in each group there will be less chance of items being missed.

Schedules

As with internal finishes measurement and for similar reasons, it is usually prudent to prepare a schedule of windows and doors before commencing measurement. Before preparing the schedule the windows should be lettered or numbered on the floor plans and the elevations. If there are several floors to the building then a system of numbering should be devised which enables windows on a particular floor to be located readily. For example, a letter could be allocated to each floor followed by a number for each window, the numbering starting at a particular point on each floor and proceeding, in say, a clockwise direction round the building. Whilst windows are usually scheduled floor by floor, there may be good reason to work by elevations or by window types or even by a combination of all three. The total number of windows entered on the schedule should be checked carefully with the total number shown on the drawings to ensure that none is missed. A further check must be made to ensure that the windows shown on the elevations tie up with those shown on the plans. Discrepancies may occur because clerestory windows or windows to mezzanine floors are sometimes shown on the elevations but not on the plans. Any differences found should be mentioned to the architect and the matter resolved before commencing measurement. The schedule should aim to set out the details of each window so that it is hardly necessary to refer to the drawings during measurement. Usually it should be possible to measure together in one group all windows of one type irrespective of their size, the wall thickness, etc.

A note should be made at the commencement of the measure-ment for each group of the numbers of windows being dealt with; care should be taken throughout that this total is accounted for in each item. A common fault of beginners is to separate the entire measurement of windows of the same type but of different sizes or in different thicknesses of walls. Such a method may give less trouble but will probably take longer. Although the grouping of windows of different sizes may require more concentration and care, proper use of the schedule will considerably simplify the work.

Doors on the plans should be lettered or numbered serially as described for windows above. External doors may be included with the window numbering, and internal blank openings included with internal doors.

Timesing

Window dimensions will probably contain a fair amount of timesing and great care is necessary to ensure that this is done correctly. As each of the subdivisions of measurement is completed, it is advisable to total the timesing of each item to ensure that it equals the number of windows being measured. If, for instance, 20 windows are being measured in a group, timesing of deductions, lintels, cills, etc; should total 20 unless a change in specification or design for a particular window requires a smaller number.

Special features

Usually any special features that definitely relate to the window (such as small canopies above or decorative brickwork underneath) will be measured with the window.

Dormer windows

In the case of dormer windows the window itself may be taken with the other windows whilst the adjustment of the roof would normally be taken with the roofs. This division will generally be found to be convenient, particularly if the roofs and windows are being measured by different persons. The opening for a dormer window may sometimes be partly in the wall and partly in the roof, in which case the wall adjustment would be made with the window measurement, in the same way as for the other window openings.

Adjustments

When measuring a group of items, the advantage of taking initially the same description for similar work and then making an adjustment for small differences often becomes evident in the measurement of windows. For example, if all the windows are glazed in clear glass except for a small proportion which have patterned glass, the simplest way of measuring may be to take all as clear glass and then, checking over carefully with the schedule, make adjustment for those that need patterned glass. If, as is not impossible, the measurement of the patterned glass to one or more windows should be missed, then if clear glass has been measured to them all, the error will be much less than if none had been measured. Similarly, when making adjustments for the openings, different decoration may be applied to the walls. If the predominating one is chosen for the deductions to all windows then, as an adjustment, the true decoration may be deducted where appropriate and the pre-dominating finish, deducted earlier, added back. Apart from minimising the effect of errors, this method also facilitates the grouping of descriptions under the same measurements.

Windows and doors

Timber, metal and UPVC windows and frames or doors and frames are enumerated and described with a dimensioned diagram. A reference to a catalogue or standard specification may remove the need to provide a diagram. The method of fixing must be shown unless this is at the discretion of the contractor. Bedding and pointing of the frame is given as a linear item. Timber window boards and cover fillets are measured as linear items, stating their cross-section dimensions and labours in the description.

Glass

The size of each pane of glass can sometimes be obtained from manufacturers' catalogues; otherwise it has to be calculated from the overall size of the window, deducting for frames and mullions, etc. Standard plain glass, when not exceeding 10 mm in thickness and not exceeding $4\,m^2$ in area, is measured superficial. The size of the panes is indicated in the description as follows:

- Not exceeding $0.15\,\mathrm{m}^2$ in area (stating the number of panes)
- 0.15 to $4\,\mathrm{m}^2$ in area

If there are more than 50 identical panes, then their number and size are stated. Raking and curved cutting to glass is deemed to be included. Non-standard panes outside the thickness and size classification given above are enumerated and the size is given in the description. If rebates for glass are over 20 mm then this is stated in the description, the rebate depth given in 10 mm stages. Adjacent panes of glass required to align with each other have to be so described. One of the rare occasions when waste is allowed for in measurement occurs in glazing: panes of irregular shape are measured as the smallest rectangle from which the pane can be obtained. Bedding the edges of glass in strips or channels is measured as a linear item.

If the glazing is double then this is stated and the glazed area is doubled for calculating the bill quantity. If hermetically sealed double glazing units or special glasses are specified then the panes are enumerated, stating their size.

Ironmongery

Each item of ironmongery, if not supplied with the window, is enumerated and described giving the nature of the background to which it is fixed. The description of ironmongery is simplified by reference to manufacturer's catalogues, being careful to give sizes, material and finish if alternatives are listed. Frequently it is convenient to include a PC sum for the supply of ironmongery and measure out the fixing. This avoids the necessity to give a full description for the ironmongery, which is impossible at the measurement stage if a selection has not been made.

Decoration

Decoration is measured as a superficial item, the measurement being taken over frames, mullions, transoms, cills and glass. The description includes the size of the panes, averaged if of more than one size, classified as follows:

- Not exceeding $0.1\,\mathrm{m}^2$
- 0.1 to $0.5\,\mathrm{m}^2$

- 0.5 to 1 m^2
- Exceeding 1 m^2

The measurement is deemed to include such items as paint on opening edges and the consequential extra frame, cutting in and work on glazing beads. Decoration that is external has to be so described. Priming only to backs of frames before fixing is measured as a linear item if not exceeding 300 mm girth.

Openings

Unless the opening has rebated reveals, the same dimensions can be used for the deduction of the wall, cavity and external and internal finishes. After making this adjustment, it is best to consider the perimeter of the opening in the order of head, jambs and cill. Precast concrete lintels are enumerated; the size, shape and reinforcement is included in the description. In situ concrete lintels are measured as isolated beams; the cubic measurement of the concrete, and the measurement of the formwork and reinforcement each have to be taken separately. Brickwork displaced by lintels is only deducted for height to the extent of full courses displaced and for depth into the wall to the extent of full half brick beds displaced. Damp proof courses to heads forming cavity trays are so described. Proprietary steel lintels are enumerated, giving the manufacturer's reference. Facework to arches is measured as a linear item, stating the number. Plaster to the reveal, if not exceeding 300 mm wide, is measured as a linear item. Decoration to the reveal is measured as the work to the adjacent wall. At the jambs, cavity closing is measured as a linear item; facework to the reveals is deemed included but vertical damp proof courses are measured. Precast concrete cills are measured in the same way as lintels with the same rules for adjustments. Facework to cills is measured as a linear item, stating whether it is set weathering and the dimensions.

Example 10

Window

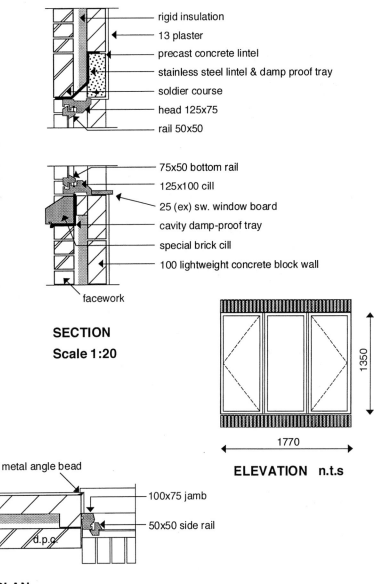

rigid insulation
13 plaster
precast concrete lintel
stainless steel lintel & damp proof tray
soldier course
head 125x75
rail 50x50

75x50 bottom rail
125x100 cill
25 (ex) sw. window board
cavity damp-proof tray
special brick cill
100 lightweight concrete block wall

facework

SECTION

Scale 1:20

1350

1770

ELEVATION   n.t.s

metal angle bead

100x75 jamb

50x50 side rail

d.p.c.

PLAN

Scale 1:20

Fig. 34

Example 10

Taking-off list		SMM reference
Softwood window		L10.1.0.1
Bed frame		L10.10
Window board		P20.4.1
Glazing		L40.1.1.2
Paint to windows		M60.2.3.1.5
Paint to window board		M60.1.0.3
Adjust wall construction:	facings	
	block	
	cavity	
	insulation	
	plaster	
	decoration	
Lintel		F31.1.1
Cavity tray		F30.16.1
Brick arch		F10.6.1
Adjust brick for arch		
Plaster to reveals		M20.1.2.1
Angle beads		M20.24
Paint to reveals		M60.1.0.1
Close cavity		F10.12.1.1
Damp proof course		F30.2.1.1
Cill		F10.15.1.3
Adjust brick for cill		
Damp proof course		F30.2.2.3

				Windows 1
	1		Softwood window size 1 770 x 1 350 mm as reference XXX , diagram figure 34, the frame fixed with four galvanised mild steel fixing cramps screwed to back of frame and built into blockwork, with easy clean hinges, with standard brass casement stay and two pins and brass casement fasteners.	L10.1.0.1 Dimensioned diagram required.
2/ 2/	1.77 1.35		Bed wood frames in gauged mortar (1:1:6) and point one side in mastic.	L10.10
	1.77		Wrot softwood window board 25 x 150 mm rebated and rounded one edge, screwed and pelleted to masonry.	P20.4.1 Ends deemed included. Note:- Ironmongery supplied with window and already fixed. If ironmongery not supplied then each item would be described and the fixing enumerated. If the specification has not been decided then a prime cost sum could be included and provision for profit and fixing would be given.

				Windows 2
			Glass	
			Width 1 770	
			Frame 2/40 = 80 Mullions 2/30 = <u>60</u> - <u>140</u> 1 630 Fixed width ÷ 3 = 543 - Side rails 2/20 <u>40</u> Open width <u>503</u>	Often manufacturers' catalogues give the glass sizes.
			Height Head 40 1 350 Cill <u>50</u> - <u>90</u> Fixed = 1 260 Top 20 Bottom <u>30</u> - <u>50</u> Open light = <u>1 210</u>	
	2		Sealed double glazed units in clear glass size 500 x 1.210 m as manufacturer's specification and glazing with sprigs and putty to wood in panes 0.15 - 4 m²	L40.3.1.1 (D4g)
	1		Ditto size 540 x 1.260 mm	
2/	1.77 <u>1.35</u>		Knot and prime, glazed wood windows panes 0.5 - 1 m² prior to fixing.	M60.2.3.1.5 The window is to be primed before fixing. The average pane size is stated.

				Windows 3
1.77			Two undercoats and one finishing coat gloss paint on primed glazed wood, windows, panes area 0.50 - 1.00 m², girth exceeding 300.	M60.2.3.1 The painting is measured over the glass. Work to opening edges of frames is deemed to be included.
1.35				
			&	
			Ditto externally	
1.77			Knot and prime only general surfaces woodwork, application on site prior to fixing.	M60.1.0.1.4 Window board
0.35				
1			Two undercoats and one finishing coat gloss paint on primed general surfaces woodwork isolated areas n.e. 0.50 m².	M60.1.0.3

				Windows 4
			<u>Opening</u> <u>& adjustment</u>	

1.77
<u>1.35</u>
		<u>Deduct</u>
		
		Wall in facings half brick thickness, vertical a.b.
		
		        &
		<u>Deduct</u>
		
		Forming cavity in hollow wall 75 mm wide including ties
		
		        &
		<u>Deduct</u>
		
		Wall insulation a.b.
		
		        &
		<u>Deduct</u>
		
		Block wall 100 mm thick a.b.
		
		        &
		<u>Deduct</u>
		
		Plaster walls width exceeding 300 mm to blockwork a.b.
		
		        &
		<u>Deduct</u>
		
		Two coats emulsion paint to plaster walls a.b.
					Descriptions for deductions for opening adjustments need only be sufficient to identify the item.

					Windows 5

<u>Head</u>

Lintel

1 770

Bearing 2/150   300

2 070

1

Precast concrete (27 N/mm² - 20 mm agg) lintel, rectangular section 2 070 x 100 x 210 mm, reinforced with one 16 mm diameter mild steel bar and build in to blockwork in gauged mortar (1:1:6).

F31.1.1

&

3 x 350 mm girth stainless steel combined lintel and cavity tray 2 070 m long and build in to blockwork in gauged mortar (1:1:6).

F30.16.1

				Windows 6
	1.77		Flat soldier arch 215 mm wide on face in facing bricks in gauged mortar (1:1:6) half brick thick, width of exposed soffit 20 mm, including pointing (in Nr 1).	F10.6.1
	2.07 0.21		Deduct Block wall 100 mm thick a.b. & Deduct Plaster walls width exceeding 300 mm to blockwork a.b.	Deduction of one complete course. This is deducted assuming a different specification used for plastering to concrete.
	2.07 1.77		Plaster walls width not exceeding 300 mm, two coats, 13 mm thick to concrete including bonding agent.	M20.1.2.1 Plastering to the exposed face and soffit of the concrete lintel.
	1.77		Galvanised perforated metal angle bead horizontal with 50 mm returns fixed to concrete with plaster dabs.	M20.24

				Windows 7
	1.77 0.10		Two coats emulsion paint plastered walls a.b.	M60.1.0.1 Not an isolated surface therefore not classed as n.e. 300mm girth.
	1.77 0.23		Deduct Wall in facings half brick thick a.b. _Jambs_	Brick displaced by arch
2/	1.35		Closing cavities 50 mm wide with 100 mm blockwork, vertical.	F10.12.1.1
2/	1.35 0.10		Damp proof course width not exceeding 225 mm, vertical, of pitch polymer bedded in guaged mortar (1:1:6).	F30.2.1.1

				Windows 8
2/	<u>1.35</u>		Plaster walls width not exceeding 300 mm, two coats 13 mm thick to blockwork.	M20.1.2.1
			&	
			Galvanised perforated metal angle bead vertical with 50 mm returns fixed to blockwork with masonry nails.	M20.24.8
			&	
			Two coats emulsion paint plastered walls a.b. X 0.10 =	M60.1.0.1
			<u>Cills</u>	
	<u>1.77</u>		Facework cill, horizontal, purpose made splayed bricks 65 x 150 x 140 throated and set projecting 50 mm and pointing to exposed faces.	F10.15.1.3

				Windows 9
1.77 0.15			Deduct Wall in facings half brick a.b.	
			$\begin{array}{rr} & 1\ 770 \\ 2/150 & 300 \\ \hline & 2\ 070 \end{array}$	
2.07 0.25			Damp proof course exceeding 225 mm wide, horizontal a.b.	F30.2.2.3 The dpc is taken beyond the jambs to give extra protection.

Example 11

Internal Door

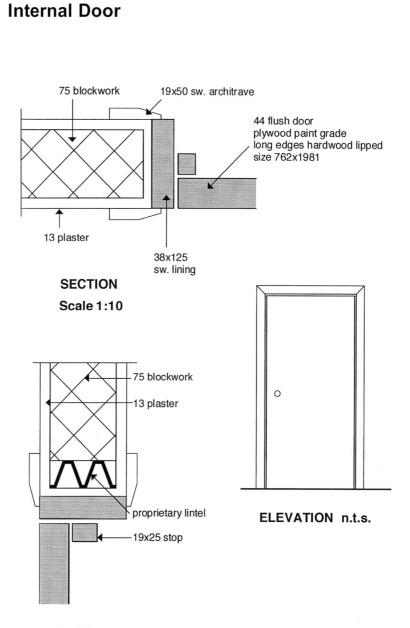

75 blockwork   19x50 sw. architrave

44 flush door
plywood paint grade
long edges hardwood lipped
size 762x1981

13 plaster

38x125
sw. lining

SECTION

Scale 1:10

75 blockwork

13 plaster

proprietary lintel

19x25 stop

PLAN

Scale 1:10

ELEVATION n.t.s.

Fig. 35

Example 11

Taking-off list	SMM reference
Door unit	L20.1.0.1
Decoration	M60.1.0.1
Ironmongery: prime cost	A52.1.1
profit	
fixing	P21.1.1
Linings	L20.7
Stops	P20.2.1
Architraves	P20.1.1
Decoration	M60.1.0.2.4
	M60.1.0.1
Adjust wall for door: block	
plaster	
decoration	
Lintel	F30.16.1.1
Adjust skirting for door	

				Internal Doors 1
	<u>1</u>		40 mm flush door 762 × 1 981 mm with painted grade plywood and hardwood lipped on long edges as diagram figure 35.	L20.1.0.1 Dimensioned diagram required
			762 1 981 <u>40</u> <u>40</u> 802 2 021	
2/	0.80 <u>2.20</u>		Knot, prime and stop, 2 undercoats and one gloss coat paint to wood general surfaces exceeding 300 mm wide.	M60.1.0.1 The door thickness is added to cover painting to edges.
			Allow the Prime Cost Sum of £_____ for ironmongery.	A52.1.1 Ironmongery supplied by Nominated Supplier.
			Add for Profit.	The Contractor must be given the opportunity to allow for profit.

					Internal Doors 2

<u>Fixing the following ironmongery to hardwood lipped softwood flush doors</u>

P21.1.1
The background is stated for fixing

  1   Pair 75 mm pressed steel butt hinges.

Sufficient detail is needed to indicate amount of labour needed for fixing

  1   Mortice lock

&

Silver anodised aluminium lever furniture.

<u>End of fixing</u>

<u>Linings</u>

Head   762
2/32   64
        826

Jamb 1 981
½/32   16
      1 997

Head

Jamb

Door size

Elevation of lining

				Internal Doors 3
			The following in No 1 door lining set tongued at angles	L20.7 The number of sets needs to be stated and an indication of the number of identical sets would be included.
	0.83		Head 32 x 115 mm wrot softwood.	
2/	2.00		Jamb 32 x 115 mm wrot softwood plugged and screwed to blockwork.	Method of fixing stated (S8)
			End of lining set	

$$\begin{array}{r} 1\,981 \\ -\quad 19 \\ \hline 1\,962 \end{array}$$

| 2/ | 0.76 1.96 | | 19 x 25 mm wrot softwood stop | P20.2.1 |

Architrave
762

$$\begin{array}{r} 2/10 \quad 20 \\ 2/75 \quad 150 \\ \hline 932 \end{array}$$

$$\begin{array}{r} 1\,981 \\ 10 \\ 75 \\ \hline 2/2\,066 \quad 4\,132 \\ \hline 5\,064 \end{array}$$

Length

Architrave

Length

Door size

				Internal Doors 4
2/	<u>5.06</u>		25 x 75 mm wrot softwood splayed and rounded architrave	P20.1.1 Measured extreme length Ends, mitres etc deemed included
			<u>Painting</u>	
2/ 2/ 2/	<u>0.83</u> <u>2.00</u> <u>0.76</u> <u>1.96</u> <u>5.06</u>		Prime only wood general surfaces n.e. 300 mm girth, prior to fixing.	M60.1.0.2.4
			<u>Paint girth</u> 115 2/19 38 2/10 20 2/75 150 2/25 <u>50</u> <u>373</u>	Girth includes architraves, linings and stops to both faces
	4.82 <u>0.37</u>		Knot, prime and stop, two undercoats and one gloss coat paint to wood general surfaces exceeding 300 mm girth.	M60.1.0.1

				Internal Doors 5
			<u>Opening</u>	
			762 1 981	
		2/32	<u> 64</u> <u> 32</u>	
			826 2 013	
	0.83 2.01		<u>Deduct</u> 75 mm thick block wall a.b.	
2/	0.83 2.01		<u>Deduct</u> Two coats plaster to wall exceeding 300 mm a.b. & <u>Deduct</u> Two coats emulsion paint general plaster surfaces a.b.	The small amount of emulsion behind architrave has not been adjusted although if finish were expensive then adjustment should be made.
			<u>Lintel</u> 826 Bearing 2/75 <u>150</u> 976	
	1		Proprietary coated galvanised steel lintel type _____ 1 000 mm long and building into blockwork	F30.16.1.1 Standard length No deduction of wall as full course not displaced

				Internal Doors 6
2/	0.93		Deduct	
			19 x 125 mm wrot softwood skirting	Adjustment for skirting here if not measured net initially
			&	
			Deduct	
			Knot, prime and stop and three coats oil paint to wood, isolated general surfaces n.e. 300 mm	
			&	
			Deduct	
			Prime only backs a.b.	
			&	
			Add	
			Two coats emulsion paint to plaster walls a.b. X 0.13 = _____	Emulsion paint deducted twice, once with skirting and once with door opening adjustment
			Adjustment for flooring in opening would be measured if not taken with finishes take-off.	

Chapter 12
Reinforced Concrete Structures

Generally

This type of work involves the measurement of a number of components comprising a combination of concrete, steel reinforcement and temporary support known as *formwork*.

The measurement of bar reinforcement, although not difficult, is beyond the scope of this book. It is, however, measured linear with an appropriate allowance for bends and hooks and then the linear measurements are converted to weight prior to billing. The calculation of the number of bars generally follows the approach used in dealing with rafters or floor joists. The distance between the first and last bars is divided by the spacing of the bars, and by adding one to the result the number of bars can be calculated. Fabric reinforcement is measured superficially to the actual area covered, the estimator allowing in the price for the loss of material due to the laps. In practice the required reinforcement is usually included on an engineer's schedules and therefore just needs to be 'extended'. Total lengths of each bar type are calculated which can then be direct billed.

The measurement of a reinforced concrete structure requires a clear and logical approach to the order of the take-off so that items are not missed. The work may be subdivided into several parts of the building then taken off floor by floor, measuring the concrete, formwork and reinforcement for each floor. In some instances it may be more practical to measure all of one type of structural component, such as columns, throughout the building. It is also advisable to measure the associated formwork after the concrete so that it is not overlooked, and in any case the same dimensions are usually applicable. Again a schedule can be of great assistance in simplifying the take-off and reducing any repetitive dimensions.

Columns

Columns should be measured between floors; in counting the number of columns, one column is taken at each grid point on every floor level. On a two storey building there would therefore be two columns at each point, one on the ground floor and one on the first floor. The concrete is measured in cubic metres. The formwork to simple columns is measured as a superficial item, taking the concrete length multiplied by the girth and stating the total number of columns. This allows the estimator to value the cost of forming the column kickers, which are not measurable, and to add an allowance per square metre to the column formwork cost.

Structural floors and roofs

Structural floors and roofs are measured to the overall dimensions to the outside edge of the columns or beams. The slab concrete is measured as a cubic item, the thickness of the slab being identified as not exceeding 150 mm, 150 to 450 mm, and exceeding 450 mm (slabs in the same class can be grouped together).

The soffit formwork is measured as a superficial item to the net area excluding beam soffits and column heads. Further information is given in the description, such as the thickness of the concrete and the support height.

Depending on the complexity of the floor plan a typical bay size could be measured and then be multiplied by the number of bays or the area could be measured overall and then the beam and column areas deducted. Formwork will also need to be measured to the slab edges and should be measured as a linear item, stating the height in accordance with SMM: E20.3.

Beams

Beams should be measured between the columns. Where these are attached to concrete slabs then the volume of concrete is added to the slab concrete (see definition rule D4(a)). This does not affect the description of the slab because it is the general thickness that is given in the description.

The formwork to the sides and soffit of the beams is measured as a superficial item using the concrete length and cross-section sizes to obtain the superficial area.

Where beam sides and floor edge coincide, the floor edge is added to the girth of beam formwork. There is no need to make any deduction from the measurements for junctions of members (see measurement rule M11).

Walls

Walls are measured in the same way as slabs and it should be noted that they include attached columns in the same way that the beams were added to the slabs. It is common practice to cast a small part of the wall with the slab. This is called a *wall kicker*; although it makes no difference to the concrete measure, it does affect the formwork. A linear item is taken for the wall kicker and measured on the centre line of the wall; the price is to include for formwork to both sides. Formwork to walls exceeding 3 m high needs to be identified and kept separate.

Reinforcement

The SMM requires that the weight of bars shall include that for bends and hooks: for each hook, one can usually add nine times the diameter of the bar rounded to the next 10 mm. If the length of the concrete is used for calculating the length of the bar, then the concrete cover must be deducted, i.e. in the case of a 12 mm bar, 110 mm must be added at each end less the amount of concrete cover.

Example 12
Concrete Frame

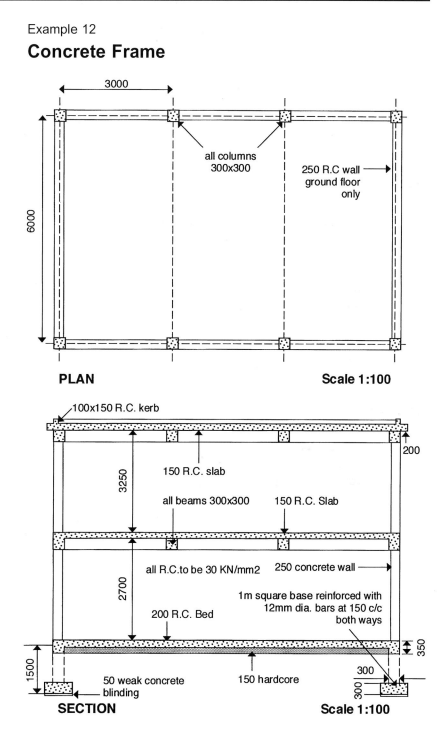

Fig. 36

Example 12

Taking-off list	SMM reference
Topsoil excavation	D20.2.1.1
Topsoil disposal	D20.8.3.2
Pit excavation	D20.2.4.3
Disposal of soil	D20.8.3.1
Adjust for backfilling	D20.9.1.1
Earthwork support	D20.7.2.2
Surface treatment	D20.13.2.3
Blinding concrete	E10.4.1.5
Concrete bases	E10.3.1
Reinforcement	E30.1.1.1
Concrete columns	E10.11
Formwork to columns	E20.15.3.1.1
Concrete bed	E10.4.2.1
Surface treatment	E41.2
Hardcore bed	D20.10.1.3.1
Surface treatment to hardcore	D20.13.2.3.1
Thickening to bed	E10.4 Note D3(c)
Formwork to edge of bed	E20.2.1.3
Working space	D20.6
Concrete wall	E10.7.2
Attached columns	See Note D6
Formwork to wall	E20.12.1
Formwork to attached columns	E20.15.2.1.2
Wall kicker formwork	E20.21
Concrete slabs	E10.5.1.1
Surface treatment	E41.2
Attached concrete beams	See Note D4(a)
Soffit formwork	E20.8
Beam formwork	E20.13.1.1
Formwork to edge of slab	E20.3.1
Concrete upstand	E10.14
Formwork to upstand	E20.4.1.2

				Concrete Frame 1
			ALL CONCRETE TO BE 30 KN/mm²	
			_Substructure	
			Depth 1 500	
			blinding 50	
			1 550	
			- topsoil 150	
			1 400	
			W B	
			6.000 3/ 3.00 9.000	
			0.300 ½ col x 2 0.300	Calculation of overall
			6.300 9.300	dimensions to outer edge of concrete foundation
			Bases	
			1 000	
			less column 300	
			2) 700	
			spread 350	
	9.30		Excavating topsoil for preservation average 150mm thick	D20.2.1.1
	6.30			
8/	1.00			
	0.35		&	
4/	0.65		Disposal, excavated material on site in spoil heap average 100m from excavation	D20.8.3.2
	0.35		x 0.15 = m³	
			&	
			Surface treatment compact bottom of excavation	D20.13.2.3

				Concrete Frame 2
8/	1.00 1.00 1.40		Excavating pit depth n.e.2m, (No 8) &	D20.2.4.3 The number of pits to be given
			Disposal of excavated material off site	D20.8.3.1
			Additional depth of earthwork support external faces bases 2/4 8 2/2 4 —— 12	
8/4/	1.00 1.40		Earthwork support n.e.2 m deep, n.e.2 m between opposing faces	D20.7.2.1
12/	1.00 0.15			To support topsoil around perimeter
8/	1.00 1.00		Weak concrete bed (1:12) n.e. 150mm thick, poured on or against face of earth x 0.05 = m^3 &	E10.4.1.5
			Reinforced concrete isolated foundation poured against face of earth x 0.30 = m^3	E10.3.1
8/	0.30 0.30 1.20		Reinforced concrete in column	E10.11 Height taken from top of pad to underside of bed

				Concrete Frame 3
8/	1.00 1.00 <u>1.05</u>		Filling to excavation exc 0.25 m thick with excavated material compacted in 150 mm layers & <u>Deduct</u> Disposal excavated material of site	D20.9.1.1 Height of excavation less depth of concrete and blinding
8/	0.30 0.30 <u>1.05</u>		<u>Deduct</u> both last	For stub columns
	Item		Disposal of surface water	
8/2/ 7/	<u>0.90</u>		Reinforcement Base 1 000 - cover 2/ 50 <u>100</u> centres 150) 900 bars = 6 +1 <u>=7</u> Mild steel bar reinforcement 12 mm diameter, straight	E30.1.1.1 Assumes a mat of straight bars in both directions in each base. To be weighed up prior to billing
8/	1.20 <u>1.20</u>		Formwork to square columns, isolated, height n.e. 1.5 m (No 8)	E20.15.3.1.1 The number to be stated

				Concrete Frame 4
			Overall length 9 300 - edge beams 2/300 <u>600</u> <u>8 700</u>	
			Overall width 6 300 - edge beams 2/300 <u>600</u> <u>5 700</u>	
	8.70 <u>5.70</u>		Filling to make up levels n.e. 250 mm thick with imported hardcore as specified x 0.15 = m^3	D20.10.1.3.1
			&	
			Surface treatment, compact surface of fill including blinding with fine material	D20.13.2.3.1
				E10.4.2.1
	9.30 6.30 <u>0.20</u>		Reinforced concrete bed 150–450 mm thick	
2/	5.70 0.30 <u>0.15</u>			D3c Thickenings are included with beds Taken between columns.
2/3/	2.70 0.30 <u>0.15</u>			

				Concrete Frame 5
2/	8.70 0.15		Surface packing to filling, vertical face	D20.12.1 Hardcore laid before edge beam cast
2/	5.70 0.15			
	9.30 6.30		Power float surface of concrete	E41.2
2/	9.30		Formwork to edge of bed, 250 – 500 mm high, plain vertical	E20.2.1.3 Includes edge beam face
2/	6.30			
12/	0.30		Deduct Last item & Add Ditto n.e. 250 mm high	Adjustment at column position, formwork taken to underside of bed.
2/3/	2.00 0.15		Working space allowance to reduced levels	D20.6.1 Taken for edge formwork and measured between bases.
2/	5.00 0.15			
8/4/	0.30 1.20		Ditto to pits	D 20.6.2 Needed for column formwork because less than 600mm space available

				Concrete Frame 6
			Superstructure	
6/	0.30		Reinforced concrete column	Two ground floor columns are
	0.30		as before	attached to wall and taken
	2.70		(ground floor)	therefore with wall concrete
8/	0.30			
	0.30			
	3.25		(first floor)	
			300/4 = 1 200	
6/	1.20		Formwork, columns, isolated,	
	2.70		square height n.e. 3 m (No 6)	
8/	1.20		Ditto height n.e. 4.50 m	
	3.25		(No 8)	
	6.30		Reinforced concrete wall,	E10.7.2
	2.40		thickness	
	0.25		150 - 450 mm	Extra concrete in projection
2/	0.30			
	0.05		(attached	
	2.40		column)	

				Concrete Frame 7

	6.30 2.40 5.70 2.40	Formwork to vertical walls (between columns)	E20.12.1
	6.30	Formwork to wall kicker, one side suspended	E20.21
2/	0.65 2.40	Formwork to attached columns, rectangular including wall end height n.e. 3.00 m (No 2)	E20.15.2.1.2

	9.30 6.30 0.15 9.70 6.70 0.15	Reinforced concrete slab n.e. 150 mm thick (roof)	E10.5.1.1

(roof)	9 000
	300
200/2	400
	9 700

	6 300
	400
	6 700

	9.30 6.30 9.70 6.70	Power float surface of concrete	E41.2

				Concrete Frame 8
²/₃/	2.70		Reinforced concrete slab as before	Note E20
	0.30			D4(a)
²/₃/	0.30			
	5.70			
	0.30			
	0.30			
	5.70		(roof)	
	0.30			
	0.30			
	6.30		(ground floor)	Columns at wall only
	0.30			measured to underside of
	0.30			beams
3/	2.70		Formwork to soffits of slab	E20.8.1.1
	5.70		n.e. 200 mm thick, height n.e. 3.00 m	
			&	
			Ditto height n.e. 4.5 m	
6/	2.70		Formwork to square attached	E20.13.1.1.2
	0.90		beams height n.e. 3.00 m	Beam support height is from
²/₄·/	5.70		(No 12)	soffit of beam
	0.90			

				Concrete Frame 9
	2.70		Formwork to square attached beams height n.e 3 m including 150 mm high edge of slab (No 8)	Note:- E20
	1.05			M12
	5.70			
	1.05			
	6.30			
	1.05		900	
			150	
			1 050	
	32.00		Formwork to soffit of projecting slab n.e. 250 mm wide height to ground n.e. 7.50 m	It is better to keep this separate from other soffits to give a clearer picture
			9 700	
			6 700	
			2/16 400	
			32 800	
			- 200/4 800	
			centre line 32 000	
	32.80		Formwork to edge of slab n.e. 250 mm high	
	30.80		Reinforced concrete upstand	E10.14
	0.10			
	0.15		32 000	
			- 800	
			31 200	
			- 100/4 400	
			30 800	
2/	30.80		Formwork to vertical sides of upstand n.e. 250 mm high	E20.4.1.2

Chapter 13
Structural Steelwork

The measurement of structural steel can generally be classified as either

- Fabricated G10.1
- Erection G10.2
- Isolated members G12.5. An isolated member is one which is not part of a frame (D7) and would cover work such as isolated beams resting on padstones.

The measurement of the framing is required to be split into two items, with the fabrication being kept separate from the erection. The item for erecting all of the steelwork giving the total weight involved and identifying whether the work is a trial erection or the permanent works (G10.2.1/2) with the unit of measurement being Tonnes.

The weight of the steel supplied to site may be greater than the calculated weight due to the manufacturing process. This is because of the rolling margin for which no allowance is made.

The supply and fabrication of the sections is generally measured by weight and the structural function needs to be stated, therefore you would have separate items for items such as columns (G10.1.1), beams (G10.1.2), bracings (G10.1.3), trusses (G10.1.8) etc.

In each of the classifications you need not have separate items for every section because there are only three weight classes stipulated for each grouping.

- Weight less than 40 kilograms per linear metre
- Weight between 40 and 100 kilograms
- Weight exceeding 100 kilograms

Therefore you would only have a maximum of three items for each major grouping.

The different steel members are joined together by an assortment of plates, brackets, angles and the like. These are called fittings and are measured separately by weight in a composite item of fittings for the whole of the structure (G10.1.10) irrespective of the member to which they are attached (measurement rule M2).

Fittings

There are two components to a connected joint in steelwork generally:

- A fitting – e.g. a cleat which is usually a short length of steel section, either an angle or channel which acts as the connecting agent and is measured as described above.
- A fixing – this could be rivets, bolts or distance pieces.

The fixing devices are given below.

- Black bolts – forged from round stock with only the threads machined and used in clearance holes about 1mm larger than the diameter of the bolt. These are deemed included in the weight of structural members.
- Turned bolts – machined on shank and under head to fit holes with a very small clearance. These are enumerated as G10.1.12 and defined in D4.
- Friction grip bolts – high tensile steel and tightened to a predetermined tension so that the load is carried by friction, nowadays often used instead of site riveting. Measured as turned bolts.
- Rivets – rarely used as a site connection as it is a very skilled operation and the noise levels may be unacceptable. No effect on measurements but the type of connections would be given in as preliminary information (P1(c)).
- Welding – an expensive type of site joint with accurate fitting needed and working platforms required. Dealt with in the same way as rivets.

Holes for bolts etc. are deemed included in the contractors prices.

Fixing bolts for isolated members are measured where they are required for fixing to other elements and would be enumerated as G20.25.

Offsite and onsite surface preparation and treatments are measured in square metres as G10.7 and G10.8.

The measurement of steel framed structures is no more difficult than any other type of construction, but, similar to other areas, the following of a logical approach is essential as it is easy to make errors in counting or to omit whole sections of the work.

One approach would be to follow an order of dealing with all work on a particular floor before commencing the next floor. By following this order there is less chance of something being missed. Within each floor you might follow the following order.

Main order
- Columns
- Main beams
- Secondary beam
- Filler beams
- Compound trusses
- Purlins
- Rails
- Braces
- Struts

Main member measurement order	*Fittings item*
• Columns (full length taken)	
	Base plates
	Splices
	Caps
	Cleats
• Beams	Shelf angles
	Cleats
	Gusset plates
	Angles
• Truss main members	
• Struts	
• Ties	

Looking at the above order, it will be seen that the columns are given priority of measure followed by the beams and so on. The fittings order extends the measure of the individual members of the first order to deal with the constituent parts. Splay cuts to members or plates are not measurable but the members would be measured the maximum length and plates measured to the nearest rectangle of metal.

The measurement of concrete casing to beams and columns would be measured under E10.10 and E10.12 and the associated formwork as E20.14 and E20.16.

Example 13
Structural Steelwork

u.c. 254 × 254 × 107 kg

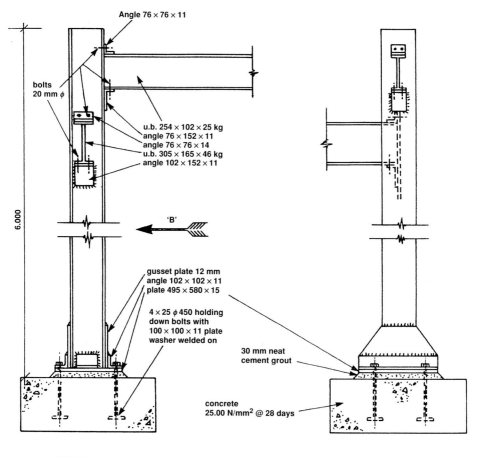

Angle 76 × 76 × 11

bolts
20 mm φ

u.b. 254 × 102 × 25 kg
angle 76 × 152 × 11
angle 76 × 76 × 14
u.b. 305 × 165 × 46 kg
angle 102 × 152 × 11

6.000

'B'

gusset plate 12 mm
angle 102 × 102 × 11
plate 495 × 580 × 15

4 × 25 φ 450 holding
down bolts with
100 × 100 × 11 plate
washer welded on

30 mm neat
cement grout

concrete
25.00 N/mm² @ 28 days

SECTION 'AA'

Scale 1 : 20

View on arrow 'B'

Fig. 37
(continued on following page)

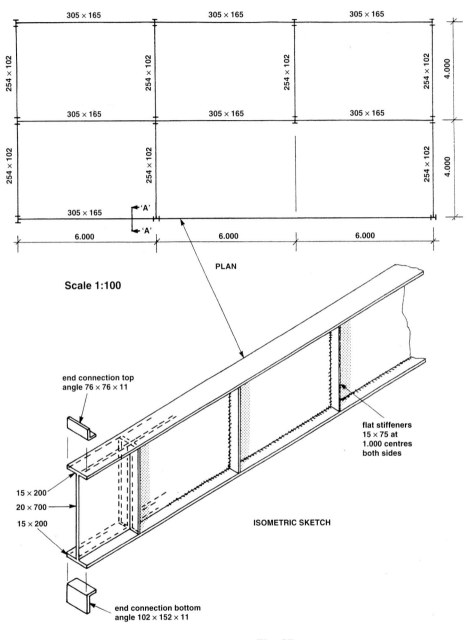

PLAN

Scale 1:100

end connection top
angle 76 × 76 × 11

flat stiffeners
15 × 75 at
1.000 centres
both sides

15 × 200
20 × 700
15 × 200

ISOMETRIC SKETCH

end connection bottom
angle 102 × 152 × 11

Fig. 37
(continued)

Example 13

Taking-off list	**SMM reference**
Framing fabrication:	
Columns	G10.1.3
Fittings: plate	G10.1.10
angle	
Holding down bolts	G10.1.11
Grout under base plate	E10.16.1
Beams	G10.1.2.2
Fittings: angle	G10.1.10
Built up girder	G10.1.8.1
Fittings: angle	G10.1.10
Erection of fabricated steel	G10.2.2

Surface treatment may be required and would normally be measured with the steel frame element

				Steelwork 1
			<u>Preamble</u> <u>notes would be</u> <u>written here</u>	Location drawings to be supplied G10.P1
			<u>Fabricated framing with</u> <u>factory welded joints and</u> <u>bolted site connections</u>	
			Column 254 x 254 Precise height 6.000 less grout 0.030 less plate <u>0.015</u> 5.955	
11/	<u>5.96</u>		Framing, fabricated in columns exceeding 100 kg/m <u>x 107kg/m =</u> kg	G10.1.3
11/	0.50 <u>0.58</u>		Framing, fabricated in fittings (Weight up the following) 15mm plate <u>x Kg/m^2 =</u> kg	G10.1.10 Fittings are grouped together and these items will be collated with all other fittings to give one bill item.
11/2/	0.58 <u>0.30</u>		12mm gusset <u>x Kg/m^2 =</u> kg	12mm plate measured rectangular (M3)
11/2/	<u>0.58</u>		102 x102 x11mm angle <u>x Kg/m =</u> kg	base
11/2/	<u>0.15</u>			

				Steelwork 2
7/2/	0.17		102x152 x11mm angle x Kg/m = kg	for 305 × 165 beam
	0.10		76 x152 x11mm angle x Kg/m = kg	for 254 × 102 beam
			(End of weighted fittings)	
11/	1		Framing, fabrication holding down bolt assemblies comprising, 4 No 25mm diameter m.s. bolts 450mm long with nut/washer and 100 x 100 x11mm plate washer welded on and setting in concrete base with all necessary f/w and wedge and pin up base plate with steel wedges.	G10.1.11
			&	
			Grout stanchion base with cement grout	E 10.16.1
			Beam 305mm length 6.000 Less 2/ 0.006 <u>0.012</u> <u>5.988</u> length 6.000 less ½ column 0.127 less steel <u>0.006</u> <u>5.867</u>	Note that the columns at end of plate girder are turned through 90 degrees and therefore one 305 beam and two 254 beams are different lengths to the others.

				Steelwork 3
6/	5.99		Framing fabrication in beams 40 -100 kg/m	G10.1.2.2
	5.87		x 46kg/m = kg	

$$
\begin{array}{lr}
 & \text{beam 254mm} \\
\text{length} & 4.000 \\
\text{less } 127/2 & \underline{0.254} \\
 & \underline{3.746} \\
\end{array}
$$

$$
\begin{array}{lr}
\text{length} & 4.000 \\
\text{less} & \underline{0.133} \\
 & \underline{3.867} \\
\end{array}
$$

5/	3.75		Ditto not exceeding 40 kg/m	G10.1.2.1
			x 25kg/m = kg	
2/	3.87			

			Fittings as before (weight up the following)	G10.1.10
7/2/	0.17		76 x 76 14mm angle	Top angle 305mm beam
			x kg/m = kg	
7/2/	0.10		76 x 76 11mm angle	254mm beam
			x kg/m = kg	
			(End of weighted fittings)	

$$
\begin{array}{lr}
 & \text{girder length} \\
 & 12.000 \\
\text{less } \tfrac{1}{2} \text{ width col x 2} & \underline{0.254} \\
 & \underline{11.746} \\
\end{array}
$$

				Steelwork 4
			Framing, fabrication in built up girder comprising plate web, flanges and stiffeners	G10.1.8.1
			Weight up the following	
2/	11.75 0.20		15mm plate x kg/m = kg	Flange
2/11/	0.70 0.08			75mm stiffeners @ 1m centres, = 12 less one
	11.75 0.70		20mm plate x kg/m = kg	
			(end of built up girder)	
			Fittings as before Weight up the following	G10.1.10
2/	0.20		76 x 76 11mm angle x kg/m= kg	
2/	0.20		102 x 152 11mm angle x kg/m = kg	
			(end of weighted fittings)	
	****		Framing, erection of permanent structure 18 x 8 x 6 metres with bolted site connections	G10.2.2 * total all weighted items = Tonnes

Chapter 14
Plumbing

Subdivision

When measuring plumbing, it is particularly important to follow a logical sequence of taking-off in order to be sure that no part is missed. Frequently, particularly on a small domestic installation, the only information shown on the drawings is the location of sanitary appliances. If this is the case then the measurement of the appliances is fairly straightforward and forms a logical start.

After the sanitary appliances have been measured, and possibly coloured in on the drawings, it is then easier to decide on a pipework layout for both wastes and supplies. Sizes of waste pipes are dictated by the size of the waste fitting from the appliance, e.g. wash basins 32 mm and baths and sinks 38 mm. Supply pipework sizes for a small installation should not be too difficult to assess. The rising main is usually in 15 mm pipework and the down feeds from the cistern in 28 or 22 mm, reducing to 15 mm for the individual feeds except for baths, which require 22 mm. Adequate isolating and drain-off valves should be included in the system to enable sections to be isolated and drained.

Gatevalves and ballvalves do not restrict the flow of water when fully open and should be used on the low pressure distribution part of the system. Some water supply companies still require an indirect system with a drinking feed to the kitchen sink taken off the rising main and the other cold feeds coming from a storage tank. The capacity of the cold water tank is given either as nominal, i.e. filled to the top edge, or actual, i.e. filled to the working water line. The requirements of water supply companies vary considerably in required capacities but this should be at least 112 litres actual for storage only, rising to 225 litres for storage and feeding a hot water system. The cold water tank may have to be raised to provide adequate pressure and flow, particularly for showers, and the roof construction may have to be strengthened to support the additional weight. The inlet to the tank is controlled by a ballvalve and the outlet should be opposite the inlet to avoid stagnation of water. An overflow pipe with twice the capacity of the inlet should be provided. An alternative installation, where permitted, is for all appliances to be fed directly from the incoming main supply, without the need for a cold water tank.

Before attempting to measure a plumbing installation, trade catalogues depicting fittings available for the specified pipework should be obtained for reference. A selection can then be made of suitable fittings for connections to various appliances. A diagrammatic layout of the plumbing when provided is often not to scale and drawn in two dimensions. When measuring from such a diagram, one has to visualise the layout in three dimensions and to relate pipe runs to the structure. This will enable realistic lengths of pipes to be measured and the correct number of bends to be taken. Sometimes, when measuring copper pipes, it is difficult to decide whether to take *made* bends (i.e. the pipe bent to form the bend) or fittings. Generally, made bends should only be taken for minor changes in direction of pipes or on short lengths. It should be remembered that long lengths of pipes with made bends may be impossible to install. For the measurement of installations without detailed information the following is suggested as a suitable order:

Sanitary appliances	(a)	Sanitary appliances including taps, traps, brackets, waste, overflow fittings, etc.
Foul drainage above ground	(b)	Wastes, overflow pipes, soil and ventilating pipes including ducts
Cold water installation	(c)	Connection to supply company's main, supply to boundary of site and meter/stopvalve pit, reinstatement of highway
	(d)	Supply in trench from boundary to building, stopvalve, rising main to cold water tank
	(e)	Branches from rising main including exterior taps and non-return valves

	(f)	Cold water tank and lid including bearers, overflow and insulation
	(g)	Cold down services.
Hot water installation	(h)	Feed from cold water tank
	(j)	Boiler, flue, controls and work in connection
	(k)	Cylinder and primary flow and return pipes
	(l)	Secondary circulation, expansion pipe and branch services.
Generally	(m)	Casings
	(n)	Chlorination, testing, etc.

Note: Insulation of pipes and builder's work in connection (e.g. chases, holes, painting, etc.) should be taken after each subdivision.

An alternative approach to measurement is to follow the flow of water from the water main, through the building to the sanitary appliances and discharging into the drains. This is a more logical approach and would probably be adopted if the layout of the whole system is shown on the drawings. The main divisions shown above could still be used but in a different order.

As far as presentation in the bill is concerned, plumbing work has to be classified under headings indicating the nature of the work. For a simple domestic type installation these would be as follows:

(a) Sanitary appliances
(b) Foul drainage above ground
(c) Cold water
(d) Hot water
(e) Sundry builder's work in connection with services
(f) Testing and commissioning the drainage or water system.

Note: For small-scale installations (c) and (d) may be combined.

The builder's work in connection section may either be billed under a heading at the end of each appropriate work section or after the installation measurement.

When taking-off, one obviously has to keep in mind these bill divisions but, by following the suggested order of measurement given earlier, the necessary sections will be automatically produced. There are several additional divisions of work for more sophisticated installations required by the SMM than those mentioned above but these are considered to be beyond the scope of this book.

Sanitary appliances

If sanitary appliances are specified fully then they are enumerated and the description should include the type, colour, size, capacity and method of fixing, including details of supports, mountings and bedding and pointing. Frequently a catalogue or BS reference is used for part of the description but care must be taken to define any alternatives available. If full details are not available then a PC sum may be included for the supply of the appliances, an item included for contractor's profit and fixing measured as enumerated items. Descriptions should make it clear whether items such as taps, traps, overflow assemblies and bath panels are included with the appliance. Small items such as towel rails, mirrors and soap dishes must not be overlooked. Any builder's work such as tile splashbacks, bearers, backboards, painting and similar items necessary for the installation should be measured at this stage.

Foul drainage above ground

Included in this section is the measurement of waste pipes, overflow pipes, antisyphonic pipes and soil and ventilating pipes. Some appliances, such as WCs, have traps built in and some, such as wash basins, have integral overflows. Traps and other pipework ancillaries are enumerated with a dimensioned description and the method of jointing stated, although cutting pipes and jointing materials are deemed to be included. Nowadays most waste and soil pipes are specified to be in plastic and their description should state whether they have ring seal or solvent welded joints and the type and spacing of pipe supports. Pipes are measured linear over fittings, and joints in the running length, i.e. jointing straight lengths of pipe together, are deemed included. The nominal size of pipes has to be given; copper and plastic are usually described by their external diameter and cast or spun iron and mild steel by their nominal bore. Straight and curved pipes have to be classified separately and in the case of the latter the radius should be stated. Fixing the pipes to special backgrounds is given as follows:

- To timber including manufactured building boards
- To masonry, which is deemed to include concrete, brick, block and stone
- To metal
- To metal-faced material
- To vulnerable materials, which are deemed to include glass, marble, mosaic, tiled finishes or similar.

Pipes laid in ducts, trenches, floor screeds and in situ concrete have to be so described. Fittings, such as bends and tees, to pipes not exceeding 65 mm diameter are enumerated and taken as extra over the largest pipe and are described as fittings with one, two or three ends, stating whether inspection doors are present. Fittings not falling within these categories are also measured extra over the largest pipe but are described with the method of jointing stated. Special joints and connections to different pipes and ancillaries are enumerated and described as extra over the pipe, stating the method of jointing. Testing the foul drainage is given as an item, giving details of the tests and any attendance required. Cutting mortices and sinkings for the installation are enumerated stating their size and the nature of the structure and any necessary making good. Linear measurements are taken for cutting chases, stating the number and sizes of pipes, the nature of the structure and any necessary making good. Holes for pipes are enumerated and grouped as not exceeding 5 mm, 55 to 110 mm, and exceeding 110 mm nominal bore (SMM P31.20). Although in some work sections holes are deemed included, those made at a later stage for services are considered to be measurable. Metal slates and collars and collars around pipes in asphalt and felt are enumerated. Painting pipes, described as painting services, is measured linear to pipes not exceeding 300 mm girth and superficial to those exceeding 300 mm girth. The measurement of overflow pipes to flushing cisterns must not be overlooked; these are measured in the same way as waste pipes.

Cold water

The measurement of the cold water installation will start invariably with the connection to the supply company's main; this would in all probability be included as a provisional sum. It is necessary to check with the company the extent of the work that they will carry out. Often included with the connection is the pipework to the meter/stopvalve pit at the boundary and making good the highway. The meter/stopvalve is required to be located on the pavement or just inside the boundary. Pipes and fittings are measured in the same way as described for waste pipes. Stopvalves, gatevalves and ballvalves are defined as pipework ancillaries and are enumerated, described and their method of jointing stated. Storage tanks are enumerated as general pipeline equipment and are described including the size and capacity. Overflows to tanks should be taken at this stage.

Insulation to pipelines is measured linear and described including the thickness of the insulation and the nominal size of the pipe. Working insulation around ancillaries is enumerated as extra over the insulation. Insulation to equipment is either measured superficial (on the surface of the insulation) or enumerated giving the overall size. In the former case, working around ancillaries is enumerated and in the latter case can be included in the item description. Excavating trenches for services not exceeding 200 mm nominal size are measured linear, stating the average depth in 250 mm stages. Earthwork support, consolidation, back-filling and disposal are deemed to be included in the trench item. Meter/stopvalve chambers and boxes are each enumerated and described. Underground ducts are measured as linear items, stating the type, nominal size, method of jointing and whether they are straight or curved. Fittings and special treatment at ends are enumerated as extra over the ducts. Timber tank bearers are measured linear and described as individual supports, giving the cross-section dimensions. The remainder of builder's work is measured as described before.

Hot water

Domestic hot water systems, apart from the pipework, have three main components – the boiler, the cylinder for storage and the cold feed storage. Suitable pipe sizes would be 28 mm for the primary flow and return between the cylinder and boiler and for the cold feed to the cylinder. The hot water distribution from the cylinder would be 28 mm, reducing to 22 mm for the vent and to 15 mm for sink or basin supplies and 22 mm for the bath. These sizes should be regarded as minima; sizes would be increased for a larger number of draw-off points. When an indirect heating circuit is included in the system then either a self-venting cylinder or a separate expansion and feed tank have to be provided. Whilst heating installations are considered to be beyond the scope of this book it is worth mentioning that a separate bill heading of low temperature hot water heating (small-scale) would have to be introduced. Boilers and cylinders are enumerated and described under the rules for equipment; the description should include, as appropriate, the type, size, pattern, rated duty, capacity, loading and method of jointing. The remainder of the work is measured as described above.

Example 14

Internal plumbing

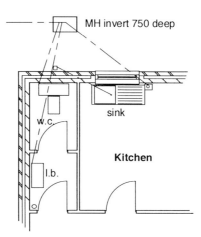

MH invert 750 deep

sink

w.c.

l.b.

Kitchen

GROUND FLOOR PLAN

soil pipe and down
service in 2 sided ply duct

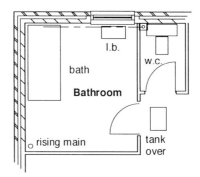

l.b.

w.c.

bath

Bathroom

rising main

tank
over

FIRST FLOOR PLAN

Scale 1:100

Fig. 38

Example 14

Taking-off list	**SMM reference**
Sanitary appliances: prime cost sum	
fixing	N13.4.1
Sink and drainer	N11.4.1
Accessories support panel	G20.12.1
Foul drainage: 110 mm pipes	R11.1.1.1.2
joint to drain	R11.2.1
bends	
32 mm traps	R11.6.8.1
40 mm ditto	
32 mm pipe	R11.1.1.1
40 mm pipe	
fittings	R11.2.3.2
joints to stack	
Overflows: pipes	
fittings	
Marking holes etc.	R11.12.1
Testing	R11.14.1
Builder's work in connection	
Cold water supply:	Y10
pipes	Y10.1.1.1
fittings	Y10.2.3.3
stopcock	Y11.8.1.1
storage tank	Y21.1.1.1
ballvalve	Y11.8.1.1
overflow	
fittings	
Marking holes etc.	Y51.1
Testing	Y51.4
Builder's work in connection	
Insulation to cold supply: pipe	Y50.1.1.1
tank	Y50.1.4.2
Painting pipes	M60.9.0.2
Tank bearers	G20.13.0.1
Backboards	P20.9.1

				Internal Plumbing 1
			<u>Sanitary Appliances</u>	
			Include the Prime Cost Sum of £_____ for Sanitary Appliances	Assuming that the specification has not yet been decided.
			Profit %	
			<u>Fixing and assembling the following including bedding taps and waste fittings in mastic and bedding and pointing at abutments with walls in flexible sealant</u>	
2/	1		White glazed vitreous china wash basins size 460 x 405 mm with pair chromium plated pillar taps, waste fitting, plug and chain and pedestal, including plugging and screwing brackets to masonry and bedding pedestal to floor and basin in mastic.	N13.4.1 As much information as possible should be given to allow the estimator to accurately assess the labour involved.

				Internal Plumbing 2
			First Floor	
<u>1</u>			White glazed vitreous china WC suite with low level 9 litre plastic cistern and ballvalve, plastic flush pipe, seat and cover, including plugging and screwing cistern brackets to masonry and screwing pan to timber and bedding in mastic.	
			Ground Floor	
<u>1</u>			Ditto but screwing pan to masonry and bedding in mastic.	The ground floor WC is fixed to the concrete floor
<u>1</u>			White reinforced acrylic bath size 1 700 x 700 mm with pair chromium plated pillar taps, overflow fitting and flexible tube, waste fitting, plug and chain, moulded front and end panel including screwing adjustable bath feet to timber, plugging and screwing wall brackets to masonry and screwing panels to timber.	

					Internal Plumbing 3
	<u>1</u>		Stainless steel double bowl, double drainer sink unit size 2 000 x 600 mm with chromium plated mixer tap, overflow fitting and flexible tube, waste fitting, plug and chain, including setting in timber base unit (measured) and fixing with screws.		N11.4.1 The timber base unit has been taken with the kitchen fittings
2/	<u>1</u>		Chromium plated toilet roll holder including plugging and screwing to masonry.		Any other items such as mirrors etc would be included here.

<u>End of fixing and assembling</u>

<u>Bath panel</u>

1 700
<u>700</u>
2 400

	2.40 <u>0.60</u>		Sawn softwood framed supports width exceeding 300 mm, 38 x 38 mm spaced at 600 mm c/c.		G20.12.1 Bath panel framing and other ancillary work for sanitary appliances taken here.

				Internal Plumbing 4
			Foul drainage above ground	
			Soil and vent pipe	
			Ground floor 2 600	
			First floor 2 600	
			Floor 250	
			5 450	
			Roof space 600	
			Above roof 450	
			1 050	
	5.45		Pipes, straight UPVC, 110 mm with ring seal joints and socket pipe clip plugged and screwed to masonry in duct.	R11.1.1.1.2 Pipes in ducts must be identified separately
	1.05 0.80 0.30		Ditto but <u>not</u> in duct	R11.1.1 Soil pipe in roof. Branch first floor WC. Ground floor WC
	1		Extra over ditto for UPVC vent terminal solvent welded.	In top of soil stack
2/	1		Extra over 110 mm UPVC pipe for joint to 100 mm clayware drain pipe socket including UPVC drain adaptor and cement and sand (1:2) joint.	R11.2.2 Joints to drain soil stack and ground floor WC

				Internal Plumbing 5
2/	1		Ditto for bent WC connector & Ditto for joint of WC outgo to socket including WC adaptor.	Special WC bend
	1		Ditto for 110 boss branch including ring seal joint.	At first floor level for WC
2/	1		32 mm polypropylene tubular swivel 'P' trap with 76 mm seal including screwed mastic joint to waste fitting and ring seal joint to MUPVC pipe.	R11.6.8.1 Washbasin traps
	1		40 mm ditto	Sink
	1		40 mm ditto bath trap with 32 mm overflow bend.	Bath

				Internal Plumbing 6
	0.60 1.50		Pipes, straight (g Flr) MUPVC 32mm (1ˢᵗ Flr) with solvent welded joints and clips plugged and screwed to masonry.	R11.1.1.1
	3.50 0.80		Ditto 40 mm	
2/	2		Extra over 32 mm MUPVC pipe for fittings two ends.	R11.2.3.2 Fittings to pipes n.e. 65 mm are described according to number of ends.
2/ 2/	1		40 mm ditto	Bath and sink
	1		Extra over 110 mm UPVC pipe for 32 mm boss branch with ring seal joint.	R11.2.4 Boss for pipe to first floor wash basin

				Internal Plumbing 7
1			Ditto 40 mm boss branch	Bath waste
1			Extra over 32 mm MUPVC pipe for joint to 50 mm cast iron drain pipe socket with UPVC socket reducer and caulking bush and cement and sand (1:2) joint.	Ground floor basin to drain joint.
1			Ditto 40 mm MUPVC pipe	Sink

				Internal Plumbing 8
			<u>Overflows to</u> <u>WC cisterns</u>	
2/	<u>1.00</u>		Pipes, straight UPVC 22 mm overflow with solvent welded joints and clips plugged and screwed to masonry.	
2/	<u>2</u>		Extra over ditto for fittings two ends. & Ditto for connection to flushing cistern including straight tank connector.	
	Item		Marking the positions of holes, mortices and chases in the structure for the foul drainage above ground.	R11.12.1
	Item		Testing and commissioning the foul drainage above ground installation.	R11.14.1

					Internal Plumbing 9
			Builder's work		
			Soil and vent pipe	P30/31. M2
1			Cutting holes in roof tiling for pipe 55 - 110 mm diameter.	H60.11
			&	
			Aluminium patent weathering slate 457 x 400 mm with moulded rubber sealing cone for 110 mm UPVC pipe handed to others for fixing.	H72.26.1.1.1
			&	
			Fixing only ditto (by roofer)	H60.10.0.0.1
1			Cutting hole in 25 mm softwood board flooring for pipe 55 - 110 mm diameter.	E20.10 P31.20.2.2

				Internal Plumbing 10
			Builders Work (Cont'd)	
2/	1		Cutting hole in 100 mm block for pipe n.e. 55 mm diameter.	P31.20.2.1 First floor wash basin and bath wastes
			End of Builders Work	
			Piped Supply systems Water Supply (Cold Water)	Y10 (M1) It is assumed that incoming main has been measured elsewhere
			Copper tubes to be to BS 2871 Part 1, half hard, table x	
			Fittings for copper tubes to be to BS 864 Part 2 compression Type A (non manipulative).	Common headings inserted here to avoid repetition
			Copper tube to be fixed with standard saddle band type clips at 1 200 mm centres fixed with brass screws.	

				Internal Plumbing 11
			Ground floor 2 600	
			First floor 2 600	
			Floor 250	
			5 450	
	5.45		Pipes, straight copper 15 mm clips plugged and screwed to masonry.	Y10.1.1.1 Rising main and branch to sink
	6.40			
			To sink 3 200	
			2 500	
			700	
			6 400	
			4 000	
			Clg 100	
			Vent to cistern 500	
			4 600	
	4.60		Ditto but clips screwed to timber	In roof
4/	1		Extra over ditto for fittings two ends	Y10.2.3.3 Bends
2/	1			
			(Sink)	
			(Sink branch)	
	1		Ditto three ends	Tee for sink

				Internal Plumbing 12
			(Ball valve) (Sink tap)	
2/	1		Extra over ditto for joint to fitting including straight connector.	Y10.2.2.1
			<u>Stopvalve</u> <u>at entry</u>	
	1		Combined high pressure screw down stop valve and drain tap as BS 1010 and 2879A, compression joints to copper tube.	Y11.8.1.1
			<u>Tank</u> <u>in roof</u>	
	1		Water storage tank to BS 4213 polythene 225 litres capacity, including lid, perforations for one 15 mm and three 22 mm pipes and backing plates (connections measured) and placing in position.	Y21.1.1.1 Water storage tank in roof

				Internal Plumbing 13
	1		Ballvalve, high pressure; BS 1968 Class A PVC float 15 mm inlet fixed to polythene tank and connection to copper pipe including straight coupling.	Y11.8.1.1
			Overflow	
	3.50		Pipes, straight UPVC 22 mm overflow with solvent welded joints and clips screwed to timber.	Tank overflow
3/	1		Extra over ditto for fittings two ends.	Bends on overflow
	1		Ditto for connection to polythene tank including straight tank connector.	Overflow connection

				Internal Plumbing 14
	Item		Marking the position of holes, mortices and chases in the structure for the cold water installation.	Y51.1
	Item		Testing and commissioning the cold water installation.	Y51.4
			Builder's work in connection with plumbing	
			Rising main	
	1		Cutting hole in 25 mm softwood board flooring for pipe not exceeding 55 mm diameter.	P31.20.2.1 Holes in plasterboard and making good plaster considered not measurable
			Sink branch	
2/	1		Cutting hole in 100 block wall for pipe not exceeding 55 mm diameter.	P31.20.2.1

				Internal Plumbing 15
	5.45 6.40		Prime and two coats gloss paint copper services, isolated surfaces not exceeding 300 mm girth.	M60.9.0.2 Painting exposed pipes
			<div align="right">Tank bearers</div>	
3/	1.50		Individual supports 50 x 100 mm sawn softwood pressure impregnated.	G20.13.0.1
			<div align="right">Tank platform</div>	
	1		Backboards etc, 1 500 x 1 250 chipboard 19 mm BS 5669 flooring grade.	P20.9.1
			<div align="right">Overflow through fascia</div>	
	1		Cutting hole in 25 mm softwood fascia for pipe not exceeding 55 mm diameter.	

				Internal Plumbing 16
			Insulation to Cold Water Supply	Y50 (M1)
4.60			Insulation 20 mm thick glass fibre sectional to 15 mm diameter copper pipes including metal bands.	Y50.1.1.1 Working around fittings is deemed included but working around ancillaries is measured.
1			Insulation 25 mm thick expanded polystyrene boards to sides and top of tank 225 litre capacity including securing with bands and cutting around pipes.	Y50.1.4.2
				Note:- The down services have been omitted from this example as it is mainly repeating items already measured.

Chapter 15
Drainage

Subdivision

The measurement of drainage may be divided into the following sections:

- Manholes or inspection chambers
- Main drain runs and fittings between manholes
- Branch drain runs and fittings between outlets and manholes
- Accessories (e.g. gullies)
- Sewer connection
- Land drains
- Testing

If the foul water and the surface water drainage are separate systems, then usually each would be measured independently using the above sequence. There may, however, be a possibility of combining manhole measurement if the construction is similar. When measuring surface water drainage, there must be liaison with the person measuring the roads and pavings in order to ascertain the position of gullies and channels. Similarly, there will have to be consultation with the person measuring the roofs in order to ascertain the position of rainwater pipes and to decide who will measure the connections if required. When measuring foul water drainage, there will have to be similar discussion with the person measuring the plumbing in order to find out the position of soil pipes and outlets and again to establish who is to measure the connections. The opportunity should be taken during these discussions to check that the drawings show drain runs leading from all service outlets.

Manholes

In most cases the position of manholes will be shown on the drawings together with invert levels. If the existing and finished ground levels adjacent to the manhole are not shown then these will have to be ascertained from the site plans, possibly by interpolation. A basic decision will have to be made as to whether excavation is to be measured from the existing ground level or, if there is excavation for other work in the vicinity of the manhole, from the reduced level. Usually the reduced level excavation takes place prior to the drain excavation and it would be correct to excavate drains from the reduced level, although one should state in the bill the method used. If there is filling adjacent to the manhole, then an assessment has to be made as to whether the filling is likely to be carried out prior to the drain excavation. When this is the case the depths could be taken from the cover level, as excavation through fill is measured as normal excavation. If the filling is likely to be carried out after the drain excavation, for example hardcore fill under roads, then the drain excavation will be measured from the original ground level.

Prior to the measurement of manholes, it is wise to prepare a schedule of similar format to that shown in Fig. 39. The schedule will enable manholes with similar plan dimensions to be spotted. Manholes with the same plan dimensions but differing depths may have their depths averaged for measurement purposes, although for excavation they must be within the same depth category. If information on the sizing of manholes is not available then the following guide may be useful:

- For inspection chambers up to 900 mm deep, the minimum internal size should be 700 mm wide × 750 mm long, allowing up to two branches on each side.
- For manholes over 900 mm and up to 3300 mm deep, the minimum size should be 750 mm wide × 1200 mm long, allowing 300 mm in the length for each 100 mm branch and 375 mm for each 150 mm branch.
- For manholes up to 2700 mm deep, a cover of size 600 × 600 mm should be provided, increasing to 600 × 900 mm for deeper manholes.
- For deep manholes an access shaft can be constructed at the top to within 2 m of the benching.
- When the depth of the manhole exceeds 900 mm, step irons should be taken at 300 mm intervals.

MANHOLE SCHEDULE						
Nr	DIAGRAM	INTL SIZE	DEPTH EXCAVATION	SIZE 150 CONC BASE	SIZE 150 RC COVER	COVER & FRAME
1		563 × 675 mm	675 Chan 50 Base 150 875	563 675 750 750 1313 × 1425	563 675 430 430 993 × 1105	457 × 457 CI BS EN 124
2 etc.						

(Manhole schedule contd.)						
	WALLS ONE BRICK			BRANCH	BUILD IN	
Nr	HEIGHT	GIRTH	CHANNEL	BENDS	ENDS	REMARKS
1	675 + 50 = 725 − 150 575	563 675 2/1238 = 2476 4/215 = 860 3336	100 mm Str 675 mm long	Two 100 mm	Four 100 mm	
2 etc.						

Fig. 39

The measurement of manholes is generally carried out as for other building work. Excavation is described in terms of pits and the number stated in the description. Precast concrete cover slabs are enumerated, but in situ slabs have to be measured in detail with separate items for formwork and reinforcement. Built-in ends of pipes, channels, benching, step irons, covers and intercepting traps are enumerated and described giving sizes. If preformed manholes are used then the excavation and associated in situ concrete work are measured in the same way but the unit itself is enumerated and fully described.

A spreadsheet could have been used to schedule the manholes. Fig. 40 is a repeat of the manhole from Fig. 39 but in a spreadsheet format, but the explanation of where the figures come from is lost without knowing the formulas in each cell. These have therefore been included at the bottom of the table for clarity only. In both cases the manhole walls have been taken as 215 mm thick, with the concrete base slab having a spread of 160 mm.

	A	B	C	D	E	F	G	H	I	J
1		MANHOLE SCHEDULE								
2		NR	Internal length	Internal width	External length	External width	Finished cover level	Invert level	Existing ground level	Excavation depth
3										
4		1	0.675	0.563	1.313	1.425	100.000	99.325	100.000	0.875
5										
6		2 . etc.								
7					=C5+(2*0.215) + 2*0.16	=D5+(2*0.215) + 2*0.16				=(15-H5)+0.20
8										

	K	L	M	N	O	P	Q	R	S
1		MANHOLE SCHEDULE							
2		Centre line bwk	Height bwk	R.c. cover 150 mm thick, length	R.c. cover 150 mm thick, width	Channel 100 dia straight, 675 long	100 dia, 3/4 branch bends	Build in ends	Cover, 450 × 450	Remarks
3									
4		3.336	0.575	1.105	0.993	1	2	4	1	
5									
6									
7		=(2*(C5+D5)) + 4*0.215	=((G5-0.15)- H5)+0.05	=C5+2*0.215	=D5+2*0.215				
8									

Fig. 40

Where a drainage layout comprises many manholes of the same construction the use of a spreadsheet can considerably reduce the extent of waste calculations, errors and time spent writing out dimensions.

Drain runs

Once the manholes have been measured, it is a comparatively simple matter to measure the main drain runs between. Firstly, a schedule should be prepared on the lines of that shown in Fig. 41. This schedule has been prepared for the drain runs in Example 15 and it could have replaced many of the waste calculations. The depth of the drain excavation will usually be the same as that for the manhole at each end, although a small adjustment may have to be made if the bed below the manhole is of different thickness to that below the drainpipe. The depths at each end of the drain run will be averaged and the depth classified in 250 mm stages. Trenches for pipes are measured as linear items; those up to 200 mm diameter can be grouped together, for those over 200 mm the size is given. Earthwork support, treating bottoms, filling and disposal are all deemed to be included with the trench item.

Bedding for pipes is classified as one of the following:

(1) Beds
(2) Beds and haunchings
(3) Beds and surrounds
(4) Vertical casings.

In the case of (2) – (4) the size of the pipe has to be given. In the case of (1) and (2) the width and thickness of the bed have to be given and in addition in (3) the thickness of the surround. In the case of (4) the size is given.

Pipes are measured as linear items measured over all pipe fittings, described as in trenches with the nominal size and method of jointing. Strictly speaking, the pipe length will be longer than the trench to allow for building in to the sides of the manhole. In practice, however, it is unlikely that measurement could be that accurate, particularly if the drawing is to a small scale. Pipe fittings such as bends are enumerated as extra over the pipe.

Branch drain runs are measured in the same way as main drain runs and again a schedule should be prepared before measurement. The important difference is that branch runs are connected to the manhole at one end only. At the other end will be either a gully or a connection to a soil pipe or similar outlet. The depth of the trench at this end will depend upon the size of the gully or rest bend but for normal circumstances could be taken as 600 mm. If bends are used, then it must be remembered that the pipe must be measured to include their length. Gullies and other accessories are enumerated and described; jointing to pipes and concrete bedding being deemed included. Back and side inlets, raising pieces and gratings should be included in the description of the gully. Usually two bends are required in a pipe coming from a gully as there may have to be a change in direction as well as an adjustment to achieve the correct fall. Circular top or two piece gullies may, however, reduce the need for direction change.

The connection to the public sewer, if carried out by the contractor, is enumerated and described. If the connection is carried out by the statutory authority, as is usually the case, then a provisional sum is included to cover the cost of the work. Care must be taken to ascertain the extent of the work done by the authority and to include for the remainder in the measured work. The work frequently involves excavation across the highway, which would require special provisions such as traffic control, lighting and reinstatement. It should also be noted that work beyond the boundary of the site has to be kept separate.

Testing the drainage system is included as an item, stating the method to be used. The extent of testing required should be ascertained from the appropriate authority.

Trench excavation depth allows for 100 mm bed and 20 mm thickness of the pipe. The length is measured between outside face of manholes.
Formation level has been assumed at 100.00, around house, and then falling away along the main runs.
Pipes are assumed to fall at least 1:40 (therefore over 30 m the drop is 750 mm and over 10 m the drop is 400 mm)
Pipes measured externally to house and up to manhole 1.

Run location	Pipe size 100 mm	Form. level at head	Invert level at head	Exc depth at head	Form. level at foot	Invert level at foot	Exc depth at foot	Ave. dp of exc	Exc tr for pipe n.e 200 mm dia		
									0.75	1.00	1.25
SVP-MH3	2.10	100.00	99.55	0.57	100.00	99.25	0.87	0.72	2.10		
Gulley-MH3	1.00	100.00	99.55	0.57	100.00	99.25	0.87	0.72	1.00		
WC to MH3	1.60	100.00	99.55	0.57	100.00	99.25	0.87	0.72	1.60		
MH3-MH2	10.00	100.00	99.25	0.87	99.75	98.85	1.02	0.945		10.00	
MH2-MH1	30.00	99.75	98.85	1.02	99.00	98.10	1.02	1.02			30.00
BED	**44.70**										
add for pipe into house	1.20										
into manholes 7 × 215	1.51										
PIPES	**47.41**							length of trenches	**4.70**	**10.00**	**30.00**

Head is the shallow end of a run, foot is the deeper end of a run.
To calculate excavation depth at head or foot is (formation level less invert) plus thickness of pipe and bed (130 mm)

Fig. 41

Example 15

Drainage

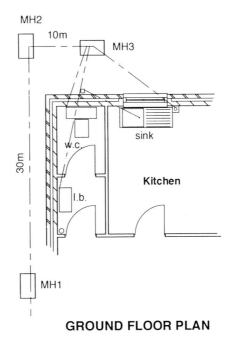

GROUND FLOOR PLAN

Fig. 42

Example 15

Taking-off list	SMM reference
Manholes	R12.11
Excavate pit	R12.11.1
Disposal of excavated material	
Earthwork support	
Compact ground	
Concrete base	R12.11.2
Brick walls	R12.11.5
Benching	R12.11.9
Main channels	R12.11.8
Branch channels	
Build in ends of pipes	R12.11.7
Concrete cover	R12.11.11
Steel cover	R12.11.11
Main pipe runs	
Excavation for trench	R12.1.1.2
Pipes	R12.8.1
Concrete bed and haunch	R12.5.1.1
Pipe fittings	R12.9.1.1
Pipe ancillaries	R12.10.1.1
Branch in building	
Excavation	
Cast iron pipe	R12.8.1.1.1
Concrete bed and surround	R12.6.1.1
Pipe fittings	
Testing	R12.17

				Drainage 1
			<u>The following in 3 no</u> <u>manholes</u>	Although SMM does not require a heading this is better for estimating purposes.
			Internal size 900 x 570 Wall 2/215 <u>430</u> <u>430</u> <u>1 330</u> <u>1 000</u>	Manhole construction is one brick wall and 150 mm concrete base with no spread.
			Depth M/H 1/2 3 Invert 900 750 Bedding 25 25 Base <u>150</u> <u>150</u> <u>1 075</u> <u>925</u>	
	1.33 1.00 <u>0.93</u>		Excavating pit n.e. 1 m deep (1 Nr) &	R12.11.1 M8
			Disposal of excavated material off site.	R12.11.1
2/	1.33 1.00 <u>1.08</u>		Excavate pit n.e. 2 m (2 Nr) & Dispose off site.	

				Drainage 2

Manholes (cont'd)

	4.66		Earthwork support n.e. 1 m deep and n.e. 2 m between faces.	R12.11.1
2/	4.66 1.08		Ditto n.e. 2 m deep.	
3/	1.33 1.00		Surface treatment, compact bottom of excavation.	R12.11.1
3/	1.33 1.00 0.15		In situ concrete (20 N) bed n.e. 150 mm thick poured on earth.	R12.11.2

```
                    900
                    570
           2/1 470 =   2 940
           4/2/½/215      860
                       3 800

       Depth  900      750
       Bedding  25       25
                925      775
    Cover slab - 100     100
                825      675
```

				Drainage 3
			Manholes (Cont'd)	
2/	3.80 0.68 3.80 0.83		One brick wall, vertical, in semi-engineering bricks, English bond in cement mortar (1:3) built fair and flush pointed one side.	R12.11.5
3/	1		Benching 900 x 570 x 190 mm average thick in concrete (20 N) finished in cement and sand (1:2) trowelled smooth to steep falls and channels.	R12.11.9
3/	3.80 0.19		One brick wall in semi-engineering bricks, English bond in cement mortar (1:3). & Deduct Ditto built fair and flush pointed one side a.b.	Adjustment below benching level

				Drainage 4
			Manholes (cont'd)	
2/	1		100 mm vitrified clay half round curved main channel 1 100 mm girth and bedding in cement mortar (1:3).	R12.11.8
	1		Ditto straight 900 mm long ditto.	
2/	1		Ditto three quarter section branch channel bends ditto.	
	4		Building in end of 100 mm pipe to one brick wall of manhole.	R12.11.7 Cutting pipes deemed included (C6)
2/	2			

				Drainage 5
			Manholes (cont'd)	
3/	1		Precast concrete (25 N) cover slab 1 330 x 1 000 x 100 mm thick reinforced with steel fabric to BS4483 ref A142 with rebated opening 600 x 450 mm finished smooth and setting in cement mortar (1:3)	R12.11.11
3/	1		Galvanised mild steel single seal cover and frame 600 x 450 mm and setting frame in cement mortar and cover in grease.	R12.11.11
				Step irons not considered necessary
			End of manholes	

					Drainage 6
				<u>Pipe runs</u>	
				Branch depth 450	
				M/H 3 <u>750</u>	
				2<u>) 1 200</u>	
				600	
				add bed and pipe <u>130</u>	
				average depth <u>730</u>	
				<u>Main run M/H 3 – M/H 2</u>	
				Invert depth 750	
				<u>900</u>	
				2 <u>)1 650</u>	
				825	
				add bed and pipe <u>120</u>	
				average depth <u>945</u>	
				<u>Main run M/H 2 – M/H 1</u>	
				Invert depth 900	
				<u>900</u>	
				2 <u>)1 800</u>	
				900	
				add bed and pipe <u>120</u>	
				average depth <u>1 020</u>	
	<u>1.60</u> <u>1.00</u> <u>2.10</u>			Excavating trenches for pipes n.e. 200mm diameter, average depth 500 – 750 mm	R12.1.1.2 If special material required for backfilling this would be given in the description.
	<u>10.00</u>			Ditto 750 mm – 1.00 m deep	
	<u>30.00</u>			Ditto 1.00 – 1.25 m deep	

				Drainage 7

	2.92
	1.22
	2.42
	10.43
	30.43
			100 mm vitrified clay pipe with flexible joints in trenches
			
			               1 000
			                215
			             1 215
			
			           10 000
			2/215 430
			           10 430
				R12.8.1.1
				600 mm is added for wall thickness for WC and soil pipe - other pipes taken to inner face of manholes.

	1.60
	1.00
	2.10
	10.00
	30.00
			In situ concrete (20 N) bed and haunching to 100 mm pipe, 425 x 100 mm section.
			
			             125
			2/150 300
			         425
				R12.5.1.1
				To outside face of manholes and measured sloping base of trench if different from plan length.
				Pipe outside diameter 125 mm.

3/	2		Extra over 100 mm pipe for bend.	R12.9.1.1 2 taken for each branch.
	1		Clay trapped gulley with 100 mm outlet, 225 mm raising piece, 225 mm high with 2 No back and side inlets, 100 x 50 mm reducing sockets and galvanised grating.	R12.10.1.1

				Drainage 8
			Branches to gulley	
	1.40		Excavating trenches a.b. n.e. 250 mm deep.	Sink GF basin
	2.20			
	1.50		50 mm cast iron pipe with spigot and socket joints in trenches in runs n.e. 3 m long (Nr 2)	R12.8.1.1.1 Jointing to gulley is deemed included (C5)
	2.30			
	1.40		In situ concrete (20 N) bed and surround to 50 mm pipe 375 × 275 mm.	R12.6.1.1
	2.20			
2/	2		Extra over 50 mm pipe for bend.	2 per run as before
			Testing	R12.17
			To take:- Drainage to sewer	Alternatively if a preformed manhole were used then it would be enumerated as R12.11.14 stating all the details.

Chapter 16
External Works

Particulars of the site

Before the measurement of external works is commenced, a visit to the site should be made to ascertain items to be included or to check the information shown on the drawings. Items which could be checked or taken include:

- Grid of levels and dimensions of the site
- Pavings etc. to be broken up
- Demolition of walls, fences, buildings, etc.
- Felling trees and grubbing up hedges
- Preservation of trees and grassed areas, etc.
- Any existing services, overhead powerlines, etc.
- Any turf that may be worth preserving.

In addition to noting these items, consideration could be given at the same time to other matters, such as access to the site, which may have to be drawn to the attention of the tenderers in the preliminaries.

Coverage

The measurement of external works can include several different aspects of work. To give an idea of the coverage a selection of items is listed below:

(1) Site clearance including:

- Removal of trees, hedges and undergrowth
- Lifting turf for reuse
- Breaking up pavings etc.
- Demolition work.

(2) Temporary works (other than those for the contractor's own convenience) including:

- Roads
- Fencing
- Maintaining existing roads.

(3) Roads, car parks, paths and paved areas including:

- Preparatory work
- Pavings
- Kerbs, edgings, channels
- Drop kerbs
- Steps and ramps
- Road markings
- Surface drainage.

(4) Fencing and walls including:

- Boundary, screen, retaining walls
- Fencing, gates
- Guard rails
- In situ planters.

(5) Outbuildings including:

- Garages
- Substations
- Gatekeepers' offices
- Bus shelters
- Canopies.

(6) Sundry furniture including:

- Bollards
- Seats and tables
- Litter, grit and refuse bins
- Cycle stands
- Prefabricated planters
- Flag poles
- Clothes driers
- Sculptures
- Signs and notices
- Lighting standards.

(7) Water features including:

- Lakes and ponds

- Ornamental and swimming pools
- Fountains.

(8) Horticultural work including:

- Cultivating, topsoil filling, subsoil drainage
- Grassed areas (seeding and turfing)
- Planting trees (tree grids and guards)
- Planting shrubs, hedges and herbaceous plants.

(9) Sports facilities including:

- Playing fields and running tracks
- Tennis courts
- Bowling greens
- Children's play equipment

(10) Maintenance including:

- Planted and grassed areas
- Playing fields

(11) External services including:

- Water, gas and electric mains
- Telephone and TV services
- Fire and heating mains
- Security systems

Whilst the above areas of work have been classified under various headings, the presentation of the work in the bill is a matter for personal preference. The SMM gives work sections for land drainage, roads and pavings, edgings, site planting, fencing and site furniture and it would be as well to follow these at least.

Site preparation

Trees and tree stumps to be removed from the site are taken as enumerated items, stating the girth measured 1.00 m above ground as 600 mm to 1.50 m, 1.5 to 3 m, and exceeding 3 m. Grubbing up roots, disposal and filling voids are deemed to be included although a description of the filling has to be given. Lifting turf for preservation is given as a superficial measurement.

Excavation

General excavation in connection with external works is measured under the same rules as for general building work. The main problems in the measurement of external works are likely to arise from excavating to reduce levels in uneven ground and from measuring irregular areas. Guidance on calculations for these items is given in Appendix 1. Provided that a proper grid of levels has been taken over the site, it should be fairly simple to find the average depth of excavation.

There may be a difficulty in giving the required depth stages for excavation when depths vary over the site. If this appears to be a problem and would mean drawing many contours representing depth changes, it is suggested that, irrespective of the required depth stages, an average depth is found for the area and this is used for the classification. The contractor should, of course, be informed that this method has been adopted. Frequently it is prudent to divide large irregular areas into sections for measurement; not only will this assist with the calculation of the area but it will also give greater accuracy in the depth measurements and classifications. Furthermore, it may be wise to make a division at the edge of an area of shallow excavation where it becomes deeper. It should be remembered that earthwork support is required to be measured only if the excavation exceeds 0.25 m in depth. Handling excavated material is generally the responsibility of the contractor but if there are any specified conditions regarding handling, such as the provision of temporary spoil heaps, then these have to be stated.

Roads and paving

Specifications for roadworks frequently refer to guides and recommendations published by the Department of Transport, the Department of Environment or the Transport and Road Research Laboratory. Concrete roads are measured as cubic items and classified as for slabs. Mechanical treatment to the surface of the concrete is given as a superficial item. Macadam roads are measured superficial, the area taken being that in contact with the base. No deductions are made for voids within the area when not exceeding 0.50 m^2. Work is described as being level and to falls, to falls and crossfalls and slopes not exceeding 15° slope, or to slopes exceeding 15°. Forming or working into shallow channels is included but linings to large channels are taken as linear items, stating the girth on face; all labours are deemed included. Gravel paving and roads are measured and classified in the same way as macadam.

Brick and block paving is also measured in a similar manner except that measurements are taken on the exposed face. The SMM includes rules for the measurement of special pavings for sport, these are measured the area in contact with the base and are again classified under the same categories mentioned above. Kerbs, edgings and channels are measured linear, with *specials* (such as corner blocks or outlets) being enumerated as extra over items.

Walling

Walling in connection with external works is again measured in accordance with the general rules although particular attention needs to be given to the measurement of any curved and battered work.

Fencing

Fencing is measured as a linear item over the posts. There are several specifications for fences and these can frequently be used to assist with descriptions. Posts or supports occurring at regular intervals are included in the description of the fencing but occasional supports such as straining posts are enumerated and described as extra over the fencing. The excavation of post holes, backfilling, earthwork support and disposal of surplus are deemed included but the size and nature of the backfilling have to be stated. Gates are enumerated and ironmongery is also enumerated separately.

Sundry furniture

Most of the items listed above under the heading of sundry furniture are enumerated and supported by a component drawing, dimensioned diagram or reference to a trade catalogue or standard specification.

External services

As far as services are concerned, pipes in trenches have to be kept separately. Special rules apply to the measurement of trenches for services; these are measured as linear items classified as for pipes not exceeding 200 mm nominal size and stating the size if over 200 mm. Average depths are stated in 250 mm stages; earthwork support, backfilling and disposal are deemed included.

Example 16

Roads and Paths

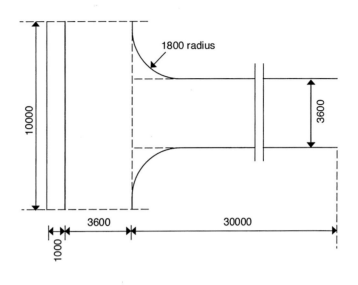

PLAN   **Scale 1:200**

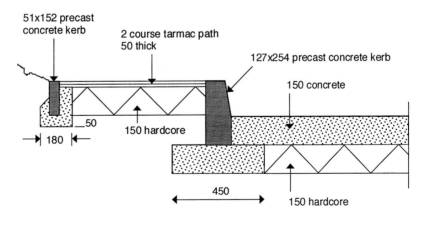

EDGE DETAIL   **Scale 1:20**

Fig. 43

Example 16

Taking-off list	SMM reference
Excavate to reduce levels	D20.2.2.1/2
Soil disposal	D20.8.3.1
Compact ground	D20.13.2.3
Earthwork support	D20.7.1.3
Curved ditto	
Filling under pavings	D20.10.1.3
Surface treatment	D20.13.2.2.1
Damp proof membrane	J40.1
Concrete bed	Q21.1
Surface treatment	E41.1.2
Reinforcement	E30.4.1
Design joints	E40.1.1.1
Concrete road kerb	Q10.2.1.2
Curved kerb	
Path kerb	
Adjustment for bell mouth area	
Tar macadam paving	Q22.2.1
Adjust soil disposal	D20.9.1.1

				Roads & Paths 1

Road length

30 000
10 000
40 000

Width	3 600	Depth
Kerb 2/127	254	254
Spread 2/162	324	150
	4 178	404
	Path	50
		150
Path 1 000		200
- Kerb 289		
711		
51		
50	101 = 812	
	Path	

10.00	Excavate to reduce levels n.e.	D20.2.2.1
0.81	0.25 m deep.	
0.20		
	&	
	Dispose excavated material off site.	D20.8.3.1

40.00	Excavate to reduce levels n.e.	D20.2.2.2
4.18	1 m deep.	The extra area bounded by curve is added later although could be included with each item.
0.40		
	&	
	Dispose off site.	

				Roads & Paths 2

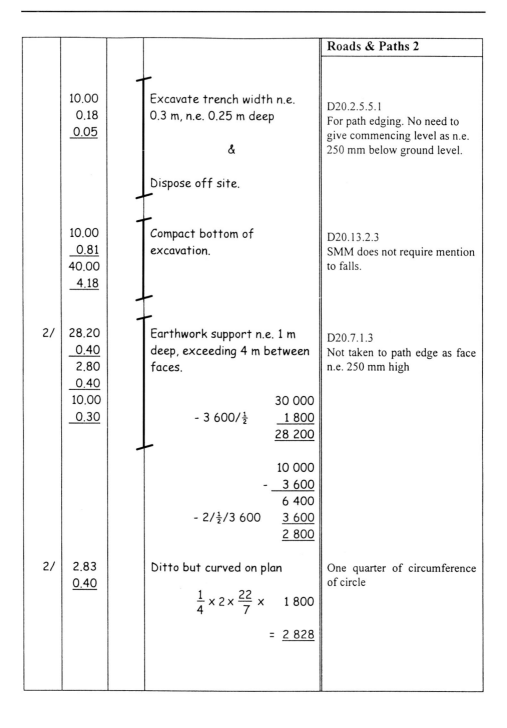

10.00 0.18 0.05	Excavate trench width n.e. 0.3 m, n.e. 0.25 m deep & Dispose off site.	D20.2.5.5.1 For path edging. No need to give commencing level as n.e. 250 mm below ground level.
10.00 0.81 40.00 4.18	Compact bottom of excavation.	D20.13.2.3 SMM does not require mention to falls.

2/ | 28.20
0.40
2.80
0.40
10.00
0.30

Earthwork support n.e. 1 m
deep, exceeding 4 m between
faces.

$$- 3\ 600/\tfrac{1}{2} \quad \begin{array}{r} 30\ 000 \\ 1\ 800 \\ \hline 28\ 200 \end{array}$$

$$- \quad \begin{array}{r} 10\ 000 \\ 3\ 600 \\ \hline 6\ 400 \end{array}$$

$$- 2/\tfrac{1}{2}/3\ 600 \quad \begin{array}{r} 3\ 600 \\ \hline 2\ 800 \end{array}$$

D20.7.1.3
Not taken to path edge as face
n.e. 250 mm high

2/ | 2.83
0.40

Ditto but curved on plan

$$\frac{1}{4} \times 2 \times \frac{22}{7} \times \quad 1\ 800$$

$$= \underline{2\ 828}$$

One quarter of circumference
of circle

				Roads & Paths 3

40.00 3.28 0.15 10.00 0.79 0.15	Filling to make up levels average 0.25 m thick with imported hardcore compacted by vibrating roller. $\qquad$ 1 000 127 79 - 206 794	D20.10.1.3

40.00 3.28 10.00 0.79	Surface treatment, compacting filling and blinding with ashes, to falls and cambers. &	D20.13.2.2.1 Because hardcore measured cube, treatment to top surface needs to be measured separately. Because of the cost falls and cambers need to be identified.
	Damp proof membrane exceeding 300 mm horizontal, waterproof building paper to BS 1521 grade B1 lapped 225 mm at joints and laid on blinded hardcore to receive concrete.	J40.1

				Roads & Paths 4
	40.00 3.60 0.15		In situ concrete bed (20 N) n.e. 150 mm thick in bays n.e. 40 m² including formwork.	Q21.1 (E10.4.1.0.1)
	40.00 3.60		Mechanical tamping concrete surface to falls.	Q21.5 (E41.1.2)
	40.00 3.60		Fabric reinforcement to BS 4483 Type A252 weighing 3.95 kg lapped 150 mm at joints.	Q21.3 (E30.4.1)
3/	3.60		Design joint in in situ concrete n.e. 150 mm deep with pre-moulded impregnated fibre board the top 15 mm filled with approved compound to BS 2499 and cutting fabric.	Q21.4 (E40.1.1.1) Assume three required

				Roads & Paths 5
2/	28.20 10.00 2.80		Precast concrete (granite aggregate) kerb to BS 7263, 127 x 254 mm bedded and jointed in cement mortar and haunched in concrete (10 N) including 450 x 150 mm concrete foundation.	Q10.2.1.0.2 Ends are deemed included (C1) but specials such as drop kerbs would be given under Q10.5
2/	2.83		Ditto curved on plan, 1 800 mm radius.	
	10.00		Ditto Fig 10, 51 x 152 mm bedded and jointed in cement mortar, haunched in concrete with 180 x 50 mm foundation a.b.	

			Roads & Paths 6	
		Adjustment for bell mouth area Area = $\frac{3}{14} \times r \times r$ r = 1 800 mm = <u>0 695</u> m²	
2/	0.70 <u>1.00</u>	Excavate to reduce level n.e. 1 m deep x 0.40 = _____ & Dispose off site x 0.40 = _____ & Compact bottom of excavation & Filling with hardcore a.b. x 0.15 = _____ & Surface treatment of hardcore a.b. & Damp proof membrane a.b.	It is best to show two dimensions to indicate that this is a square metre item.

				Roads & Paths 7

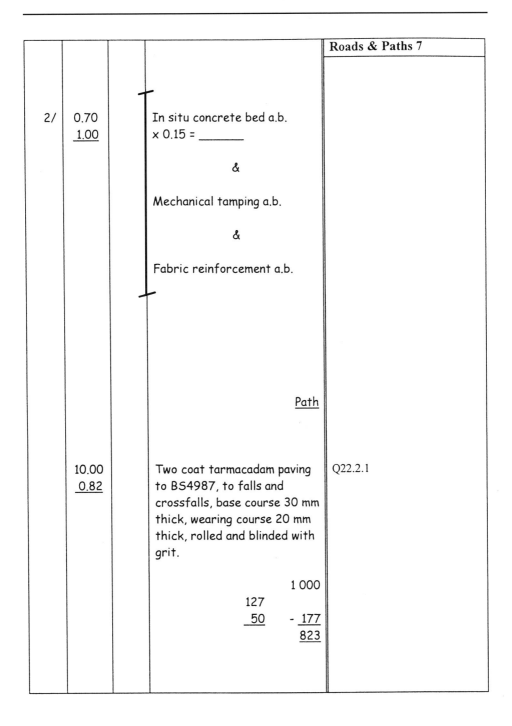

2/	0.70	In situ concrete bed a.b.	
	1.00	x 0.15 = _____	

&

Mechanical tamping a.b.

&

Fabric reinforcement a.b.

<u>Path</u>

	10.00	Two coat tarmacadam paving	Q22.2.1
	0.82	to BS4987, to falls and	

crossfalls, base course 30 mm
thick, wearing course 20 mm
thick, rolled and blinded with
grit.

```
                    1 000
           127
            50      - 177
                      823
```

				Roads & Paths 8

	10.00		Deduct	
	0.16			
	0.15		Dispose soil off site.	
	59.20			
	0.16		&	
	0.25			
2/	2.83			
	0.16		Add	
	0.25			
			Backfill to excavation n.e.	D20.9.1.1
			0.25 mm thick with excavated	Adjustments for filling behind
			material arising from the	kerb and deduction of off site
			excavations compacted in	disposal previously measured.
			150 mm layer.	

Chapter 17

Demolitions, Alterations and Renovations

Generally

As the work covered by this chapter relates entirely to existing buildings, it should be noted that all work on, or in, or immediately underneath work existing prior to the current project must be kept in a separate section of the bill.

Traditionally, alteration works have been referred to as *spot items*, *works on site* or *works to be priced on site* (which incorporated demolition, repair and alteration work). The SMM now subdivides the work as follows:

- Demolition
- Alterations – spot items
- Repairing/renovating concrete/brick/block/stone
- Chemical damp proof courses for existing walls
- Repairing/renovating metal and timber
- Fungus/beetle eradication.

In addition to the above the SMM includes a section of additional rules for work to existing buildings, covering work groups H, J, K, L, M, R and Y.

The main points to remember when measuring this type of work are firstly to identify carefully the exact location of the work, secondly to describe clearly what has to be done and thirdly to state which new items are included as opposed to being measured separately. To illustrate the last point: when describing cutting a new opening, state whether or not the new lintel and flooring in the opening are to be included in the price. Often it is advisable to measure and describe the items when on site so that one is aware of the existing conditions. Usually any locations given should relate to the existing building rather than to the proposed layout, thus enabling estimators to locate them easily when pricing on site.

Surplus materials arising from their removal become the property of the contractor, who is responsible for their disposal. If the employer wishes to retain any of the surplus materials or reuse them in the works, then special mention has to be made of this in the bill. If it is felt that any of the surplus materials may be of use or value, then the contractor could be requested to price their removal with a separate amount for credit, with a proviso that the employer reserves the right to retain the materials, allowing the contractor the credited amount, if any. This is probably best catered for in the bill by including two cash columns against the items, one being headed 'credits'.

Shoring and scaffolding incidental to demolition, repair and alteration work are deemed to be included together with making good any damage caused by the erection or removal. However, shoring of structures that are not being demolished is measured as an item. Any temporary roofs and screens required are included as items with a dimensioned description, stating details of weather and dustproofing if required. Details must be given of any toxic or special waste likely to be encountered during the work.

Preambles

Inclusion of special preambles in the bill for this type of work is necessary in order to advise the contractor of any restrictions or other matters such as phasing of the work. Furthermore, requirements common to several descriptions can be included as preambles in order to reduce the length of the descriptions. For example, 'cutting openings is to include for quoining up jambs with brickwork toothed and bonded in gauged mortar to the existing work.'

Demolitions

Demolitions are divided into three main categories as follows:

- Demolishing all structures
- Demolishing individual structures
- Demolishing parts of structures (excluding cutting openings and removing finishings etc.).

All demolition work is measured as items and sufficient information must be given in the description to enable identification and the levels down to which the demolition is to be carried out. If any particular methods of demolition are required or any restrictions are to be imposed, for example, on the use of explosives, then these have to be mentioned. Particulars have to be given of any services to be diverted, maintained or sealed off temporarily. Making good the parts of the structure remaining and any materials to be stored for reuse or retained by the employer have to be detailed. Breaking up parts of the building below the ground is usually left to be included as extra over the excavation items in the measured work.

Alterations – spot items

This work includes removing fittings, fixtures, plumbing, engineering installations, finishings and coverings. Also included are cutting openings and recesses, cutting back projections, cutting to reduce thicknesses and filling openings. None of the work included in this section should be in preparation for bonding or renewal, which is covered by separate rules mentioned below. All work is measured as items and described giving details or a dimensioned description. Care must be taken to ensure that the item can be identified and that details are given of any making good of the structure or finish required and of extending the finishes. If any new work is included, then its description must be equal to that required in the SMM for the measurement of new similar work.

Repairing/renovating concrete/brick/block/stone

Cutting out and replacing concrete, brickwork, blockwork, or stonework may be measured as a superficial, linear or enumerated item as appropriate; each with a dimensioned description. In the case of concrete, details should be given of the concrete, stating whether it is plain, reinforced or gun applied, together with any treatment to the reinforcement and bonding. Formwork and making good to match the existing are deemed to be included. In the case of brickwork, blockwork and stonework, details of the materials should be given, together with the method of bonding.

Repointing brickwork, blockwork or stonework is measured superficially with information in the description of such matters as size and depth of raking out, type and mix of pointing, bond and size of bricks, etc. Rules are also included in the SMM for resin or cement impregnation and injection, removing stains, cleaning surfaces, inserting wall ties, re-dressing to form new profiles and artificial weathering.

Chemical damp proof courses for existing walls

This work is measured as a linear item, describing the method and type of chemicals, the centres for drilling holes, details of finishes to be removed and whether to brick, block or stone. Making good to holes and finishes after injection is deemed to be included.

Repairing/renovating metal and timber

This work, which also includes refurbishment, is measured as a superficial, linear or enumerated item as appropriate. The work is described by means of a dimensioned description and must give all the necessary information including identification of the work to be explored, prepared and executed, together with any associated works required.

Fungus/beetle eradication

The treatment of existing timber is measured superficial, linear or enumerated as appropriate and accompanied by a dimensioned description. A description of the treatment is given together with identification of the work to be explored, prepared and treated.

Additional rules relating to work to existing buildings

As mentioned above, these additional rules apply only to certain sections of the SMM and are used when preparing for bonding to or renewal of the item. With the exception of services, the rules include such items as bonding and jointing new to existing; stripping off, removing or taking down; raking and curved cutting and cutting holes. As far as services are concerned, the rules include such items as breaking into existing, jointing to new, stripping out, provision of temporary services and testing.

Example 17
Spot Items

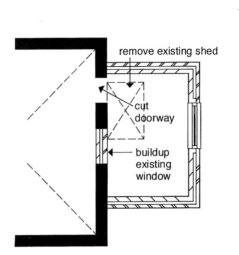

PLAN   **Scale 1:100**

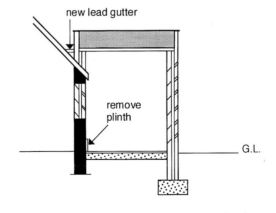

SECTION   **Scale 1:100**

Fig. 44

Example 17

Taking-off list	SMM reference
Take down buildings	C10.2.1.1.3
Break up floor	C10.3
Filling openings	C20.8
Forming openings	C20.5
Concrete lintel	F31
Cut back plinth	C20.6
Work to eaves	C20.2
Roof covering adjustment	C20.4

				Spot Items 1
Item			Pull down to ground level existing shed adjacent to east wall of kitchen approximately 1 900 x 1 300 and 2 100 high, include making good remaining brickwork and facings.	C10.2.1.1.3 Indicate location, size and making good.
1.90 1.30			Break up and clear away concrete floor and hardcore under (assume 150 mm concrete and 150 mm hardcore). ALTERNATIVELY	An assumed thickness needs to be taken unless existing drawings are available. If subsequently this is not correct then an adjustment to the price may be necessary.
Item			Break up and clear away concrete floor and hardcore under approximately 2.5 m² (assume 150 mm concrete and 150 mm hardcore).	C10.3

				Spot Items 2

<u>Filling openings</u>

<u>Item</u>

Take out and clear away window 1 200 x 1 200 mm in east wall of kitchen complete with frame and stone sill under. Fill opening with skin of half brick in commons and skin of 100 mm insulating blocks with cavity and ties width to suit wall thickness. Cutting and bonding to jambs and pinning up to soffit. Plaster both faces and adapt and extend wall tiling on one side, make good all work to match existing.

C20.8
As much information as possible needs to be given to allow the estimator to assess the work involved.

A description such as this meets the requirements of the SMM.
However, where a large number of windows of the same size are involved then it may be best to measure separate items to assist the contractor's pricing.

				Spot Items 3

<u>Form openings</u>

<u>Item</u>		Cut opening 850 x 2 050 mm in 1½ brick wall adjoining last window for new door opening, cut out for and insert new lintel (measured separately), face up jambs, make out flooring and skirting to new opening and make good plaster up to new linings.	C20.5

	<u>Door</u>
	750
frame 2/50	100
bearing 2/150	300
	1 150

<u>1</u>	327 x 140 mm precast concrete lintel 1 150 mm long reinforced with and including 3 no 12 mm mild steel bars.	F31 Alternatively the lintel could have been measured with the new door.

New door: to take	A reminder to ensure this work not missed.

				Spot Items 4
Item			Cut back projecting brick plinth 25 x 300 mm on the east elevation for a length of approximately 4 400 mm, including making good both ends up to new wall.	C20.6
Item			Take down eaves gutter and fascia for a length of approximately 4 400 mm including making good both ends of fascia and providing stop ends to gutter.	C20.2 It would be necessary to check on site that no additional work to down pipes was necessary.
Item			Strip roof coverings and cut back projecting rafters for a length of about 4 400 mm, prepare $1\frac{1}{2}$ brick wall for raising, including making good roof up to new gutter.	C20.4
			Gutter and eaves work: to take	Reminder as before

Chapter 18
Preliminaries

Generally

The drafting of the preliminaries, preambles and PC and provisional sums sections of the bill, which are not usually derived from the dimensions, is discussed in this chapter. These sections are often left to a late stage in the bill preparation, as reference to the draft bill of measured items may be necessary. The specification will form the basis of these sections; if items have been *run through* in the specification as they are measured, then those remaining may have to be either included or, if the specification is being provided with the tender documents, referred to in these parts of the bill. It should be noted, however, that specification items that relate to method and quality of workmanship are usually excluded from the bill unless they affect cost. With the introduction of the Common Arrangement for Specification and Bills of Quantities the drafting of preliminaries and preambles became simpler and the risk of something being missed is greatly reduced.

Preliminaries and general conditions section

It must be remembered that the bill of quantities has to set out all the circumstances and conditions that may affect the contractor's tender. Items that are of a general nature and do not necessarily relate to the quantity of permanent work are set out in the preliminaries section for pricing by the estimator as method-related charges, divided into time-related and fixed charges (SMM A, D1 and D2). A *time-related charge* is one that is considered to be proportional not to the quantity of the item but to the length of time taken to execute the work. A *fixed charge* is one that is considered to be proportional neither to the quantity of work nor to the time taken. The various parts of this section are discussed below.

Location drawings

Three types of drawings have to accompany the bill of quantities, as follows:

- Block plan – to identify the site and outline of the building in relation to the town plan or other wider context.

- Site plan – to locate the position of the buildings in relation to setting out points, means of access and general layout of the site.

- Plans, sections and elevations – to show the position occupied by the various spaces in a building and the general construction and location of the principal elements.

Each work section in the bill must be introduced with a statement giving the scope and location of the work unless it is shown on the above or on additional drawings sent to tenderers or in the specification information. The SMM, at the commencement of each work section, lays down the exact nature of the information to be given.

Project particulars (SMM A10–13)

This section sets the scene of the proposed contract and full particulars of the project, including the location; details of the employer and consultants, together with a list of drawings, are given. A full description of the site and details and dimensions of the new buildings and how they are to be constructed must be set out by the provision of either drawn information or measurement descriptions (SMM A.M1). Details should be given about existing buildings if their presence is likely to affect the cost of the new work, for instance a mediaeval listed building or a multi-storey car park with continuous traffic problems.

Contract particulars (SMM A20)

Particulars of the contract have to be given but if a standard published form of contract is being used then the conditions need not be repeated in full in the bill provided that the form of contract together with any relevant amendments is named and the numbers and titles of the clauses are listed. The decision as to which clauses require pricing is left to the estimator. If the conditions of contract are not standard then the conditions must be included in full in the bill or alternatively presented as a separate volume with the tender documents and listed in the bill for pricing as above.

Modification of contract conditions

Special clauses may sometimes be added to the conditions of a standard form of contract and these together with any modifications to clauses should be detailed in the bill. Such amendments must also be made to the actual contract that is to be signed by the employer and successful contractor, as it is not advisable to rely on a clause in the bill alone even though it may be one of the contract documents. Clauses 2.2.1 and 14.1 of the JCT form emphasise that the bills of quantities are concerned only with the quality and quantity of work, and the former clause makes it clear that the bill cannot override the contract.

Employers' requirements and contractors' cost items (SMM A30–44)

Broadly speaking this part of the preliminaries section contains two types of items: firstly, those that have a separate identifiable cost such as insurance, site facilities and various fees payable, and, secondly, those costs that arise from a particular method of carrying out the work. The latter would include fixed costs for items such as providing the plant and bringing it to and removing it from the site, and time-related costs for items such as maintaining the plant on site or for providing supervision for the works. Employers' requirements, which have to be described as fixed and time related, include the provision of site offices, fences and name boards, insurances and other fees and charges. A list of suggested items to be included under this heading is given in the SMM (A36). Limitations such as working hours and sequence of work required by the employer have also to be given. Contractors' general cost items, which again have to be described as fixed or time related, include for example supervisory staff, site accommodation, facilities such as water, power and lighting, plant and temporary works. Also included is an item for general attendance on sub-contractors, which is dealt with more fully under PC sums below.

PC sums (SMM A51–52)

The extent to which these are defined in the form of contract should be examined carefully as some supplementary definition may be necessary. Separate sections (A51–52) are included in the SMM relating to PC sums for nominated subcontractors and suppliers. The utilisation of the nomination process under JCT 98 is not as common as in the past, however, when PC sums are included, the contractor must be given the opportunity to add for profit to each sum. There must be no doubt as to what cash discount for prompt payment the contractor can expect in relation to PC items in the bill. If a 5% cash discount is allowed then the contractor may decide to add little or no profit, but if there is no discount at all then no doubt an addition will be made to cover profit. It will be seen, therefore, that tenders could be affected substantially by any doubt about the discount. The conditions of contract in JCT 98 state that PC sums for work to be carried out on site by 'nominated sub-contractors' referred to in clause 35 do not include a cash discount but PC sums for goods to be supplied by 'nominated suppliers', referred to in clause 36, and fixed by the contractor are to include a cash discount of 5%.

This demonstrates how important it is to distinguish between PC sums for nominated suppliers and those for nominated sub-contractors. It is common practice nowadays to have a separate section in the bill for PC and provisional sums, with each type listed under headings. This has the advantages of facilitating adjustment in the final account and providing for easy identification of the various types of items. On the other hand the true cost of each work section is not obtained simply by taking the total of that section from the priced bill, but requires the appropriate PC sums to be added.

Carriage of the goods to site should always be included in the PC sum for nominated suppliers, as the estimator cannot be expected to price for transport from an unknown source. This shows that the quantity surveyor has to be careful to read the small print in a quotation to ascertain whether anything has been excluded that should be drawn to the tenderer's attention in the bill or rectified by obtaining a revised quotation.

Work covered by PC sums for nominated subcontractors must be described as fully as possible in order that the contractor can price preliminary items taking this work into account. General attendance, which provides for the use by the subcontractor of facilities that have already been provided by the main contractor, is defined in the SMM (A42.C3). General attendance is included as an item in the preliminaries. Special attendance (SMM A51.1.3), for example, special scaffolding beyond the main scaffolding already in place should, whenever possible, be described fully; otherwise it has to be covered by a provisional sum in the bill.

Provisional sums (SMM A53–54)

These are included for work for which there is insufficient information available for proper description. Provisional sums may also be included to cover possible expenditure on items that may be required but for which there is no information available at tender stage. Here the SMM states that provisional sums may be for either undefined or defined work (i.e. work that can be described fairly fully). For the latter the contractor is expected to have taken the work into account when pricing other sections, such as indirect costs in preliminaries. The effect of this is that when the actual sum expended is ascertained for the final account no adjustment to other prices in the bill is made. On the other hand, if the work is undefined then other prices such as, for example, plant items, may have to be adjusted when the work is valued for the final account.

Works that have to be carried out by local authorities or statutory undertakings should be included as provisional sums (SMM A53). When such bodies quote for general contract work this is included in the bill as a PC item; for example, an electricity company quoting for an electrical installation as opposed to mere connection to the mains. A further provisional sum should be included to cover the cost of telephone calls made on behalf of the client, if appropriate. A contingency sum is usually included in the bill as a provisional sum. This sum covers unforeseen additional work. The amount is deducted from the contract sum in the final account. Alternatively, sums can be included to cover specific items of risk to the employer, the true extent of which cannot be fully established until the works are carried out.

A further subdivision of provisional sums is necessary when the JCT Intermediate Contract is used. Under this contract, there are no nominated subcontractors or suppliers; they become domestic subcontractors and suppliers named by the architect. The contractors pricing the bill may be provided with the chosen subcontractors' tenders and invited to include the sums, together with any additions the main contractor wishes to make, in the main tender. Alternatively the main contractors are invited to include provisional sums to be set against subcontractors' tenders in due course. When this is the case, these provisional sums should be identified separately.

Specification preambles

Preambles can be defined as general descriptions of workmanship and materials relating to the work sections which may affect price but are probably better excluded from the item descriptions. For example it may be necessary to describe cement used in making mortar for brickwork to be BS 12 but it would not be practicable for this to be included in every description of brickwork involving the use of mortar. Commonly, preamble items form a separate section in the bill, away from the work sections, although many surveyors prefer to insert them at the start of each appropriate work section. Preambles may also include methods of measurement adopted for items that are not included in or are contrary to the SMM. If a full specification has been prepared by the architect using the common arrangement and has a similar layout to the bill then this can be referred to instead of writing preambles in full, but it is important to ensure that the documents are compatible and can be cross-referenced readily. Of course, it is necessary to supply a copy of the specification with the tender documents and to make it a contract document.

Daywork (SMM A55)

Daywork is work for which the contractor is paid on the basis of the cost of labour, materials, and plant plus an agreed percentage for overheads and profit. Payment in this way is usually reserved for items that cannot be measured and priced in the normal way. Daywork payments may arise in contract variations for items such as breaking up unexpected obstructions in the excavations or for adjustment of provisional sums. To enable a percentage rate for overheads and profit to be established at tender stage, it is necessary to include items in the bill and, in order to obtain competitive rates, these items should be included in such a way that they affect the total amount of the tender.

One method of achieving this is to include a number of hours for both labourers and craftsmen which the estimator prices at an hourly rate and extends into the cash columns of the bill. Some surveyors split these hours into various crafts; of course, it needs to be made clear that these rates must include for all charges in connection with the employment of labour, plus profit and overheads, etc. Sums can also be included in the bill for plant and materials; the tenderer is invited to add a percentage that represents what will be added to the net cost of these items for incidental costs, overheads and profit, should the need for daywork arise.

Chapter 19
Bill Preparation

As described in Chapter 3 the widespread utilisation of computerised systems has to a great extent made the labour intensive manual processing of dimensions redundant and therefore only brief mention of such processes is made here. Preparation of the bill itself is addressed in more detail because (although it too is now often part of an automatic process), as with setting down dimensions, it is important to understand the process involved in properly framing descriptions and the structure of the bill; thus when computerised systems are used the surveyor can ensure that the bill is complete and correct and that the necessary level of information is being provided to the contractor for pricing.

Abstracting

In splitting up the building into its constituent parts for measurement, the taker-off follows a systematic method but there has to be a repetition of the same description in different parts of the dimensions. Moreover, in the same group of dimensions there will be items that will appear in different sections of the bill, which is usually divided according to the sections of the SMM.

Under the traditional method of processing dimensions, the function of the abstract was to collect similar items together and to classify them primarily into the SMM sections and subsequently according to certain accepted rules of order and arrangement, to put them in a suitable sequence for writing the bill. The abstract was prepared by copying the descriptions and squared dimensions from the taking-off on to abstract paper in a tabulated form as nearly as practicable in the order of the bill.

The need for an abstract was eliminated by the use of the cut and shuffle system: the principles behind the abstract were retained, but the order was effected by noting each item on a separate slip of paper and sorting them into the correct order, as described below.

Cutting and shuffling

In the cut and shuffle system once the squaring has been caried out and checked, each sheet is copied before any cutting is done and the sheets to be cut must be listed, so that there is a record of the number of slips under each taking-off section. The sheets are then cut and sorted into piles, each of which represents a section of the final bill. Wasted slips, whether blanks or nilled, are filed together and not destroyed.

Each section is then taken in turn and sorted into bill order. Slave slips are collected and attached to their master slips (deduction slips being put at the end) and each bill section is tagged together. The slips are then ready for editing. A check should be made at this stage by listing the number of cut slips in each section of the bill, adding the wasted slips and checking the total with the number of slips in the original uncut copy.

Although in this system the abstract is saved, the editing work carried out at this stage is probably heavier than in the traditional method, where much coordination was done by the biller. The editor must check the descriptions and order of items and insert group headings on separate sheets. The bill section headings and preambles may be written on normal bill paper, and items to be written short are given a distinctive mark. The editor will do all the work in red or another distinctive colour and will insert 'm^3', 'm^2', 'm' or 'Nr,' on the master slips, as required to be printed in the bill, leaving room for the quantity to be inserted later. If extra items are found necessary in editing, they should be written on a spare blank slip and also on the blank copy with the same reference number. Similarly if a sheet is nilled, it should be nilled on the copy.

The collection of totals on each master and its slave sheets is carried out and the result is entered in the lower part of the waste column of the master sheet. The total in metres will then be written against the unit of measurement and entered on to the master slip. Both processes must, of course, be checked. It does seem an advantage to have all the resulting slips (sometimes thousands) serially numbered, and if there is any cross-referencing to be done, e.g. with spot items, this numbering is useful as the references can be inserted before typing. The bill is then ready for typing, after which the bill is processed as described below.

Purpose of the bill

Before considering the preparation of the bill, one should bear in mind the main purposes of the document. First and foremost the bill is used to assist contractors with the preparation of an estimate for tendering. The priced bill, which becomes a contract document, provides a basis for the valuation of varied work and of work completed for stage payments during the contract. In addition, the bill should be of use to the contractor in the organisation of the work and to the surveyor to provide historic cost information.

Procedure

Traditionally, the writing of the bill involved, in theory, copying out the descriptions and quantities from the abstract or sorted slips in the form of a schedule or list on paper ruled with cash columns for pricing, but in practice it was a good deal more than this. The taker-off may describe an item briefly if it is in common use and leave it to the billing stage for the full description to be compiled. In fact the term *working-up* originated from the task of working-up or expanding the brief taking-off descriptions into proper bill items.

Furthermore, where several measurers are working on the same project, their descriptions must be coordinated. They cannot constantly be asking each other for the exact wording of the description for a particular item, and so the different taker-off will sometimes describe the same work slightly differently. The biller must therefore understand what is being written and see where different descriptions approach each other so closely that in effect they mean the same thing. Conversely, the biller must be able to detect subtle differences in descriptions which may affect the price of an item.

Division into sections

The bill of quantities has been traditionally divided up into work sections similar to those contained in the SMM. With the introduction of SMM7, the bill commonly reflects the main common arrangement sections of the rules. In certain countries it is customary to invite separate tenders for each trade, but in the UK, unless a form of management contracting has been adopted, it is usual to invite tenders from a general contractor only. Where separate tenders are invited, the need for separate bills for each trade is obvious, but even where tenders are obtained from a general contractor the division into work packages is of assistance in pricing and simplifies the estimator's task in respect of the parts of the work to be sublet. Moreover, it is this first step in the sub-division of the bill which enables items to be easily traced. The sections into which the SMM is divided generally align with sub-contractors' work and any departures, such as precast concrete bollards being contained in furniture/equipment, are soon recognised. Work to existing buildings, work outside the curtilage of the site and work to be subsequently removed have to be billed in separate sections (SMM GR7). Traditionally, substructure or work up to and including damp proof course and external works are also kept in separate sections. These indications of location should assist the estimator in pricing and furthermore facilitate the valuation of, and measurement of possible amendments to, this work at a later stage. When a project consists of several buildings, it is often desirable to provide a separate section in the bill for each one. Alternatively, as suggested in the MC, a multi-column analysis on the page facing the descriptions and quantities can be used if the construction of the blocks is similar.

Elemental bills

For certain reasons it may be desirable to produce an elemental bill, in which the main divisions are design elements or constituent parts of the building, e.g. foundations, floor construction, windows, etc., irrespective of the work sections of the SMM. The main purpose of such a bill is to assist a standardised system of cost analysis, which may be adopted particularly where buildings of a similar nature are to be repeated. The taking-off is done by elements and each element forms a separate section in the bill, the normal order of work sections being maintained within each element. Whilst in theory this type of bill should make estimating more accurate because the items are related to a particular part of the building, contractors who sublet work may have difficulty in collecting appropriate items together. Furthermore, estimators may find similar items occurring in different elements and have the additional problem of relating the prices for these separated items.

To overcome these difficulties some surveyors offer tenderers the bill in either traditional or elemental formats, computer sortation making this particularly easy.

General principles

It must be remembered that in most cases the bill is a contract document and therefore the descriptions must be absolutely clear and usually without abbreviations, other than those in common use. The biller must have sufficient knowledge of construction to understand the descriptions; if any appear to be vague or ambiguous the taker-off should be consulted as to the exact meaning and an amendment made if necessary. The draft bill should be written on one side of the paper only, the back being used for items that may have to be inserted at a later stage, such items being clearly referenced to the exact position into which they are to be typed. After each item is written into the draft bill the corresponding item on the abstract or cut and shuffle slip should be run through with a diagonal line. It is then clear which items have been billed and which remain to be dealt with.

Order of items in the bill

The order of items in the bill is a matter for personal preference but the following general order is suggested:

(1) Work sections as contained in the SMM, although locational sections such as substructure or external works may be required.

(2) (a) Subdivisions of work sections as contained in the SMM
     (b) Subdivisions as required by the SMM, such as external paintwork
     (c) Subdivisions of different types of materials such as different mixes of concrete or different types of brick.

(3) Within each subdivision in stage (2) the order of cubic, square, linear and enumerated items.

(4) Labour-only items should precede labour and material items within the subdivisions in (3) above.

(5) Items within each subdivision in (3) and (4) above are placed in order of value, least expensive first.

(6) Preambles and PC and provisional sums usually form a separate bill, although in certain circumstances they may be contained in the appropriate work section.

Format of the bill

The bill for each work section should be commenced on a new sheet. The ruling of the paper and a typical heading are shown below:

BILL 4 SUPERSTRUCTURE

MASONRY

			£			
	BRICK WALLING					
	Common Brickwork in cement, lime and sand mortar (1:1:6)					
A	Walls, half brick, vertical	97	m²			
B	Chimney stacks, two bricks, vertical	3	m²			

The first (left-hand) column is for the item number or letter references and the binding margin. Sometimes an additional vertical line is ruled to separate the binding margin from the reference column. The main wide column is for headings, subheadings and descriptions. Frequently, as illustrated above, an unruled portion is left at the top of each page for main headings such as work sections. The next columns contain the quantity of the item and the unit of measurement. Some surveyors prefer to enter the unit of measurement first so that the quantity is adjacent to the rate in the next column to facilitate the extension of the cash sum; others prefer to see the measurement figure separated from the cash figure so that there can be no confusion between the two. The last three columns are left blank for use by the estimator, who enters the rate for the item per unit of measurement and the cash extension. Sometimes, however, a cash amount is entered by the surveyor in the last two columns, for example when entering PC and provisional sums.

Referencing items

It is essential that every item in the bill can be referred to and found easily. Some surveyors number each item consecutively through the whole bill, whilst others prefer to use the page number followed by a letter entered alphabetically against each item on the page. The letters 'I' and 'O' are usually omitted to avoid any possible confusion with numerals. If the number of items on a page exceeds the number of letters in the alphabet, then the lettering continues at this stage by using 'AA', 'AB', 'AC', etc. The latter method has the disadvantage that it cannot be done finally until the bill is typed, as the draft will have a different number of items per page. On the other hand, serial numbering has the disadvantage that items inserted or deleted at the last moment disrupt the numbering sequence; the addition of a letter to the number of an inserted item overcomes this problem – or an item deleted may have 'not used' entered against it.

Units of measurement

Great care must be taken, particularly when there is a change, to enter the correct unit of measurement. For example, an item measured as a cubic item and indicated as a superficial item in the bill may result in a considerable difference in price. Even if the error is detected and queried by a tenderer, it may involve the issue of an addendum to the bill – which should be avoided if possible. To assist in avoiding mistakes, particularly when items are inserted, it is preferable to enter the unit of measurement against each item rather than using two dots to indicate repetition.

Order of sizes

Sizes or dimensions in descriptions should always be in the order: length, width, height. Sometimes the width or front to back dimension of, say, a cupboard, is referred to as its 'depth'. If there is likely to be any doubt the dimensions should be identified. For example:

> sink base unit 1000 mm long × 600 mm wide × 900 mm high overall . . .

Use of headings

Generous use of headings will not only help estimators to find their way about the bill but may also reduce the length of descriptions. Headings generally fall into one of four categories:

(1)   Work section headings, such as 'Masonry'
(2)   Subsection headings, such as 'Brick walling'
(3)   Headings that partly describe a group of items to follow, such as 'Common brickwork in cement mortar (1:3)'
(4)   Subdivisions required by the SMM, such as 'Cold water'.

Work section headings are often repeated in the top right-hand corner of each page. Other headings should be repeated when a new page is started. This enables it to be seen at once what is being dealt with, thus avoiding the need to search back a few pages. A new heading usually terminates a previous heading; where a new heading is not used, or if there is likely to be any doubt, 'end of . . .' should be entered in the appropriate place in the bill.

Writing short

It is often convenient for pricing purposes to keep items together in the bill which would, following the normal rules, be separated. For example, it is probably easier for an estimator to price the fittings on a gutter whilst dealing with the particular gutter although this means that enumerated items have to be dealt with in the middle of linear items. The method of entering these items is known as *writing short*; it avoids breaking completely the sequence of linear items, as follows:

D	112 mm UPVC straight half round gutter fixed to timber with standard brackets	125	m			
E	Extra for stop end	9	nr			
F	Extra for angle	7	nr			

As the example shows, written short items are inset so that the main items stand out. In the written short items the words 'extra for' have been repeated instead of using 'ditto', thus avoiding any possible confusion with the use of 'ditto' in the main item. Items that are written short automatically refer to the main item under which they are written, so there is no need to refer back. The SMM may require items that are often written short to be described as 'extra over'; this means that the main item has been measured over the written short item. In other words, the estimator is to allow for the extra cost of the item only as it displaces the main item. 'Extra over' is sometimes omitted from the description when the item is written short, but it would be safer to mention that this has been done in a preamble. Writing short is usually confined to enumerated items although some surveyors write short linear items which relate closely to superficial ones. For example, a linear item of rounded edge tiles may be written short to the wall tiling. There must, of course, be a close relationship between the two items and it must be considered more convenient for pricing when presented in this way.

Unit of billing

The general unit of billing for other than enumerated items is the metre. The dimensions and their collections having been taken to two decimal places, the bill entries are rounded up or down to the nearest metre. The main exceptions are steel bar reinforcement and structural steel, which are billed in tonnes to two decimal places.

Framing of descriptions

The framing of descriptions was introduced in Chapter 2. Care must be taken to leave no doubt as to their meaning, particularly as the bill is a legal document. The opening phrase of a description should indicate the principal part of the item so that the estimator knows immediately what it is about.

The word 'approved' should be avoided if it leaves any doubt about the quality of the item. Its sole justification is to point out to the contractor that the architect's approval should be obtained for the article to be used where there are several alternatives all costing about the same. If some particular material is described and the words 'or other approved' are added to the description, a gambling element is at once introduced which is directly contrary to the purpose of the bill of quantities. One estimator may decide to price for a less expensive alternative in the hope that the architect will approve it, whilst another may price the material specified. The words 'or other equal and approved' are sometimes used by public authorities after the specification of a proprietary article to avoid the criticism that a monopoly situation has been created, favouring a particular firm.

Long-winded descriptions should also be avoided and any superfluous words omitted. The surveyor must cultivate the difficult art of making descriptions concise and easily understood and at the same time omitting nothing that is essential to the estimator for pricing. In the following, for example, the words underlined could be left out:

Small 19 × 19 mm wrot softwood cover fillet planted on around bathroom door frame
50 × 175 mm sawn softwood in floor joists spiked to timber wall plates.

It should be noted that in the above, if fixing is not specified, it is at the discretion of the contractor. In these cases, planting or spiking would normally be omitted as it is the most economical way of fixing. Furthermore, the location of an item is not normally required.

Consistency in spelling is important if only to make the bill look like a workmanlike document. Often there are alternatives for spelling technical words such as 'lintel' or 'lintol' and 'cill' or 'sill'. An easy solution to this problem is to establish a *house rule* that spelling should be as shown in the SMM. Misspelt words are another problem and this, whilst often unfairly attributed to typist's errors, can only be eradicated by careful reading over; *spell checkers* in word processing systems can assist, particularly if technical terms are added to the memory.

Furthermore, consistency in language is important, as minor differences in phraseology may cause an estimator trouble in deciding if the meaning is intended to be different. For example:

(a) raking cutting <u>to</u> existing Code 4 milled lead
(b) curved cutting <u>on</u> existing Code 4 milled lead

In this case there is no actual difference to be inferred from the word 'on'. In some cases, however, descriptions that achieve the same final result may have different values. For example:

(a) formwork to rectangular mortice not exceeding 500 mm girth in concrete depth not exceeding 250 mm
(b) cutting mortices in in situ concrete 200×200 mm depth 100–200 mm

In the first example the mortice is formed by inserting formwork prior to pouring the concrete; in the second example the mortice is cut in the hardened concrete.

Totalling pages

There are two ways in which the surveyor may indicate how the cash totals on each page are to be dealt with. Firstly, the total may be carried over to be added to the next page and so on to the end of the section. The foot of the page is completed as follows:

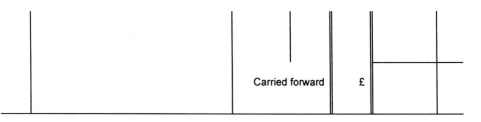

Carried forward   £

and the top of the following page as:

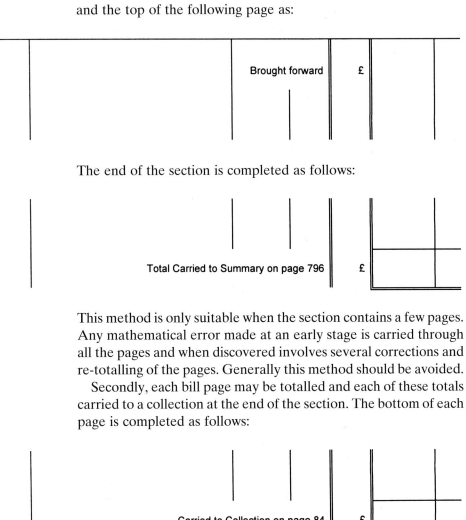

The end of the section is completed as follows:

This method is only suitable when the section contains a few pages. Any mathematical error made at an early stage is carried through all the pages and when discovered involves several corrections and re-totalling of the pages. Generally this method should be avoided.

Secondly, each bill page may be totalled and each of these totals carried to a collection at the end of the section. The bottom of each page is completed as follows:

The collection is on the last page of the section as follows:

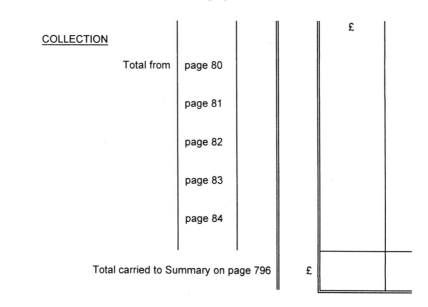

COLLECTION

		£
Total from	page 80	
	page 81	
	page 82	
	page 83	
	page 84	
Total carried to Summary on page 796	£	

Summary

At the end of the bill a summary is prepared for insertion of the totals from the collections of each section. To enable the total cost of any section to be established easily, it may be advisable to enter the title of each section as follows:

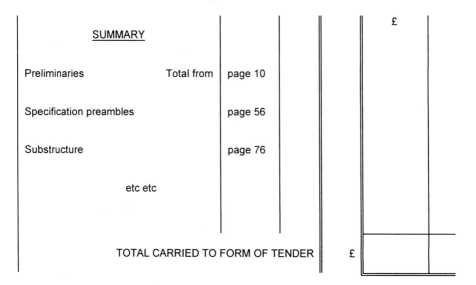

SUMMARY

			£
Preliminaries	Total from	page 10	
Specification preambles		page 56	
Substructure		page 76	
	etc etc		
TOTAL CARRIED TO FORM OF TENDER		£	

Sometimes a summary is made at the end of the work sections comprising superstructure and the total of this is carried to a general summary at the end of the bill, thus reducing the length of the latter. At the end of the final summary, provision may be made for the addition of items such as insurance and water for works, the price of which may be dependent on the total cost of the work. All such items are described fully in the preliminaries and a note made against them that they are to be priced in the summary.

The process of checking

When the bill has been produced it must be checked very carefully. Each item as it is checked should be ticked in red on the left of the draft bill and the original item run through in red. The points to be looked for are as follows:

(1)   In each item:

(a)   Correctness of figures
(b)   Correctness of units of measurement, especially that changes from one unit to another, e.g. from cube to square, square to linear, etc., are properly indicated
(c)   Correctness of descriptions, especially any figures contained therein, making it quite clear whether these are in metres or millimetres.

(2)   Generally:

(a)   Sections and subsections of the bill headed properly and headings carried over where necessary
(b)   Order of the items
(c)   Proper provision for page totals and their carrying forward or collection.

These points may appear fairly obvious, but it is extremely easy in the rush to finish a bill, which is not unusual, for some inaccuracy to be overlooked. A fascia intended to be 40 mm may be billed as 45 mm, especially if the figures are poorly written. Special care is necessary when checking items transferred or inserted from the back of the sheet, to ensure that such items do not disturb the sequence of descriptions.

Numbering pages and items

It is important to see that the pages of the draft bill are numbered in sequence before the checking is begun, as a page lost that would be noticed when checking might not be spotted afterwards. If a later section has to be checked before an earlier one, the pages should be numbered temporarily. When the bill is complete, one should fasten it all together and look through it, making sure that all the pages are present and in the correct sequence and that all items are referenced either by letters or sequential numbering, the latter being best done at this stage.

Typed proof

A word may be said about reading over the proof prepared by a typist – a further stage in checking the final document. The reading should be done from the handwritten draft, as the typist may have misread a word or figure and this interpretation, if read out, may be accepted as correct by the person looking at the draft. Besides actually comparing the words and figures and ensuring that all the figures are in the correct unit, all of which are important, special attention should be given to the repetition of headings. This is important as the pages will almost certainly end at different places in the proof to where they end in the draft. It is easy for a typist to miss out an item altogether if similar ones follow each other, particularly if the quantities are the same, and particular care should be taken when reading over these items. If the bill has been typed from cut and shuffle slips, then the checking process mentioned at the start of this chapter avoids the need to read over the proof.

General final check

The whole of the dimensions and abstract or slips should be examined carefully to ensure that all calculations have been ticked and all items have been run through. Any item not dealt with should be drawn to the attention of the taker-off or other person concerned. An opportunity should be provided for the taker-off to look through the complete bill as intentions may have been misinterpreted and errors in descriptions may stand out which might otherwise pass unnoticed. It should be noted that taking-off is not checked unless, perhaps, a junior is measuring for the first time. Therefore it is advisable to make a check of the main quantities in the bill by making comparisons between items or by making approximate calculations, possibly on the following lines:

- Check the total cube of excavation with the total cube of disposal of soil.
- Compare the topsoil excavation and the area of floor finishes per storey, making allowances for the external walls and projection outside the building.
- Calculate the total floor area of the building on all floors inside the external walls and compare this with the total area of the floor finishes and beds.
- Make an approximate measurement of the external walls, deducting all openings, and compare with the total area of walling in the bill.
- Count the number of doors and windows and compare with the total numbers of each in the bill. Ironmongery quantities also could be compared with the totals.
- Check the number and type of sanitary appliances with those billed.
- Measure the approximate area of roof tiling or flat roofing and check with the bill.
- Make an approximate check on any items that lend themselves to this, e.g. length of copings and eaves gutters, number of cupboards or other fittings, area of external pavings, number of manholes, etc.
- Compare the painting items with the corresponding items in other sections as far as possible, e.g. take the total area of all wall or ceiling plaster compared with the areas of decoration on plaster.

These checks must not be expected to produce an exact comparison. Their principal purpose is to ensure that the quantities are not wildly out through some serious mistake; if the figures are reasonably near those in the bill the checks will have served their purpose. Even if, owing to pressure of time, this check is left until the bill has been sent out to tenderers, errors found may be notified to them and thus taken into account.

Cover and contents

The front cover of the bill should as a minimum have the title and location of the work, the date and surveyor's name and address. The name of the employer is also often included. A sheet should be included at the front of the bill, listing the contents; this is particularly important in the case of a large bill.

Other bill types

Other titles for bills which have a special purpose may be encountered and are mentioned here for reference purposes. *Reduction bills* are special bills prepared when the tender figure is too high and a reduction in price is obtained by altering the work in some way. The bill may contain omissions and additions to the original. *Addenda* bills contain details of work required which is additional to the original design, determined after completion of the original bill.

Specialist bills may be required to obtain tenders for specialist work, e.g. electrical installation, which is to form nominated subcontract work. These bills should contain the full preliminaries section of the main bill and should be accompanied by the necessary drawings.

Bill of approximate quantities

This bill, also known as a *provisional bill*, is used when there is insufficient information available to prepare an accurate bill of quantities. Although the quantities are approximate, the descriptions should be correct. The bill is used to obtain rates for items from tenderers; as the production information becomes available, a *substitution bill* can be prepared and priced using the approximate bill as a basis.

Schedule of prices or rates

For smaller contracts without a bill and where the specification and drawings are the contract documents, a schedule of descriptions, without quantities, can be prepared. The contractor is asked to enter rates against the items, which are then used for valuing variations to the design. The schedule seldom attempts to be comprehensive and the rates are often unreliable. This method has also been used for contractor selection when drawn information is so scanty that even a bill of approximate quantities is impracticable.

The following figures 45, 46 and 47 show how the dimensions for some substructure measurement are extended and checked, and then transferred to the abstract and billed. It is important that figures can be traced from the dimensions through to the bill and that, during the final account stages, figures in a bill can be substantiated or easily amended if variations occur.

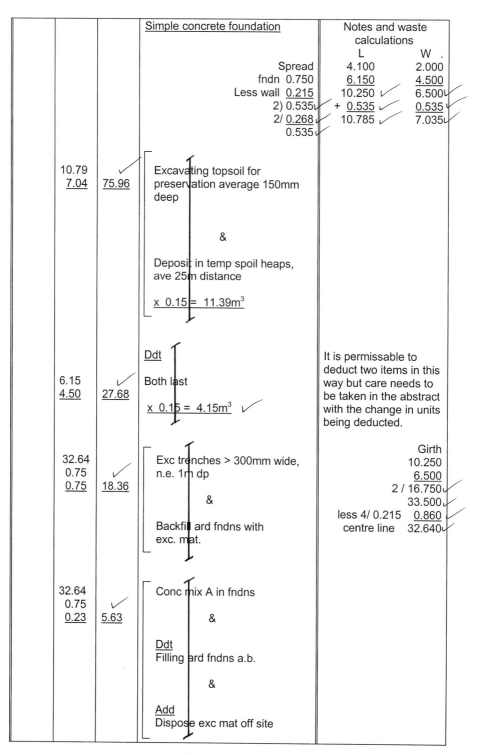

			Simple concrete foundation	Notes and waste calculations

<table>
<tr><td colspan="3"></td><td></td><td align="right">L</td><td align="right">W .</td></tr>
<tr><td colspan="3"></td><td align="right">Spread</td><td align="right">4.100</td><td align="right">2.000</td></tr>
<tr><td colspan="3"></td><td align="right">fndn 0.750</td><td align="right">6.150</td><td align="right">4.500</td></tr>
<tr><td colspan="3"></td><td align="right">Less wall 0.215</td><td align="right">10.250 ✓</td><td align="right">6.500✓</td></tr>
<tr><td colspan="3"></td><td align="right">2) 0.535✓</td><td align="right">+ 0.535 ✓</td><td align="right">0.535 ✓</td></tr>
<tr><td colspan="3"></td><td align="right">2/ 0.268</td><td align="right">10.785 ✓</td><td align="right">7.035✓</td></tr>
<tr><td colspan="3"></td><td align="right">0.535 ✓</td><td></td><td></td></tr>
</table>

10.79 7.04	✓ 75.96		Excavating topsoil for preservation average 150mm deep
			&
			Deposit in temp spoil heaps, ave 25m distance
			x 0.15 = 11.39m³
			Ddt
6.15 4.50	✓ 27.68		Both last
			x 0.15 = 4.15m³ ✓
32.64 0.75 0.75	✓ 18.36		Exc trenches > 300mm wide, n.e. 1m dp
			&
			Backfill ard fndns with exc. mat.
32.64 0.75 0.23	✓ 5.63		Conc mix A in fndns
			&
			Ddt Filling ard fndns a.b.
			&
			Add Dispose exc mat off site

Notes and waste calculations:

It is permissable to deduct two items in this way but care needs to be taken in the abstract with the change in units being deducted.

Girth
10.250
6.500
2 / 16.750✓
33.500 ✓
less 4/ 0.215   0.860 ✓
centre line   32.640✓

Fig. 45   Taking-off

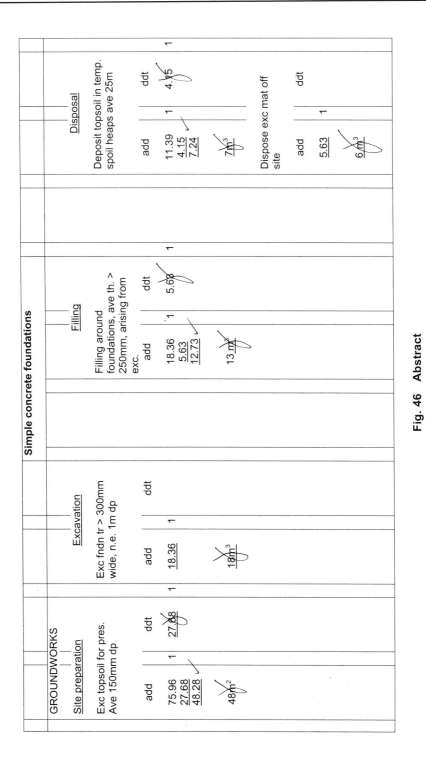

Fig. 46 Abstract

	Description of Work	Quantity	Unit	Rate	£	p
	GROUNDWORK					
A	Excavating topsoil for preservation, average 150mm deep.	48	m^2			
B	Excavating trenches over 300mm wide, not exceeding 1m deep.	18	m^3			
C	Filling to excavations, average thickness greater than 250mm, arising from the excavations.	13	m^3			
D	Disposal excavated material off site.	6	m^3			
E	Disposal excavated material on site in temporary spoil heaps not exceeding 25m from excavation	7	m^3			
	To collection			£		

Fig. 47

Appendix 1
Mathematical Formulae and Applied Mensuration

Formulae for areas (A) of plane figures

Square $A = S \times S$

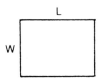

Rectangle $A = L \times W$

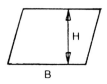

Parallelogram $A = B \times H$

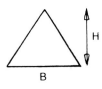

Triangle $A = \dfrac{B \times H}{2}$

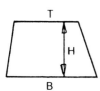

Trapezoid

$$A = \frac{(B + T) \times H}{2}$$

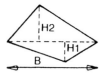

Trapezium

$$A = \frac{B \times H1 + B \times H2}{2}$$

Circle

$$A = \pi \times r \times r$$
$$\text{or} \quad A = 0.7854 \times D \times D \ \left(\text{Note}\ \frac{\pi}{4} = 0.7854\right)$$

$$(\text{Circumference} = \pi \times D \text{ or } 2\pi r)$$

Sector of circle

$$A = \frac{r \times a}{2} \text{ or } A = \frac{q}{360} \ \pi r^2$$

$$\left(\text{Note length of arc} = \text{angle} \frac{q}{360} \times 2\pi r\right)$$

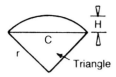

Segment of circle

$A = S - T$

Where S = area of sector
T = area of triangle

or approximately

$$A = \tfrac{1}{2} \times \left(\frac{H \times H \times H}{C} \right) + \left(\tfrac{2}{3} C \times H \right)$$

Where H = rise
C = chord

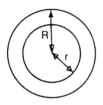

Annulus

$A = \pi\,(R + r) \times (R - r)$

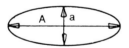

Ellipse

$A = 0.7854 \times (A \times a)$

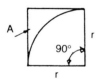

Bell mouth
(at road junction)

$A = 0.2146 \times r \times r$

Regular polygons

Pentagon	(5 sides)	$A = S \times S \times 1.720$
Hexagon	(6 sides)	$A = S \times S \times 2.598$
Heptagon	(7 sides)	$A = S \times S \times 3.634$
Octagon	(8 sides)	$A = S \times S \times 4.828$
Nonagon	(9 sides)	$A = S \times S \times 6.182$
Decagon	(10 sides)	$A = S \times S \times 7.694$

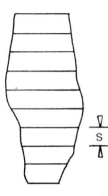

Irregular figures

Divide figure into trapezoids by equidistant parallel lines (ordinates or offsets)

$$A = S \times \left(\frac{P}{2} + Q \right)$$

(Where S = distance between ordinates
P = sum of first and last ordinate
Q = sum of intermediate ordinates)

or

Simpson's Rule (must be even number of trapezoids)

$$A = \frac{S}{3} \times (P + 4 \times Z + 2 \times Y)$$

(Where Z = sum of even intermediate ordinates
Y = sum of odd intermediate ordinates)

Formulae for surface areas (SA) and volume (V) of solids

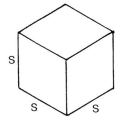

Cube

$$SA = 6 \times S \times S$$
$$V = S \times S \times S$$

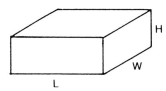

Rectangular prism

$$SA = 2(L \times W) + 2(L \times H) + 2(W \times H)$$
$$V = L \times W \times H$$

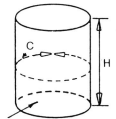

B = area of base

Cylinder

$$SA = (C \times H) + (2 \times B)$$
$$V = B \times H$$

Where $B =$ area of base $(\pi \times r \times r)$
$C =$ circumference $(\pi \times D)$

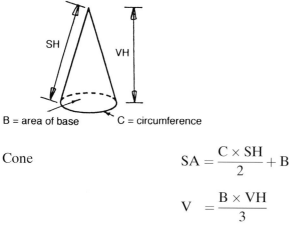

B = area of base   C = circumference

Cone

$$SA = \frac{C \times SH}{2} + B$$

$$V = \frac{B \times VH}{3}$$

Where B $=$ area of base $(\pi \times r \times r)$
 C $=$ circumference of base
 SH $=$ slope height

b = area of top

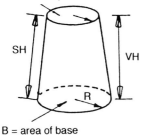

B = area of base

Frustum of cone

$$SA = \pi \times SH \times (r + R) + b + B$$

$$V = \frac{VH}{3}(\pi r^2 + \pi R^2 + \sqrt{B \times b})$$

Where B $=$ area of base
 b $=$ area of top
 R $=$ radius at base
 r $=$ radius at top
 SH $=$ slope height

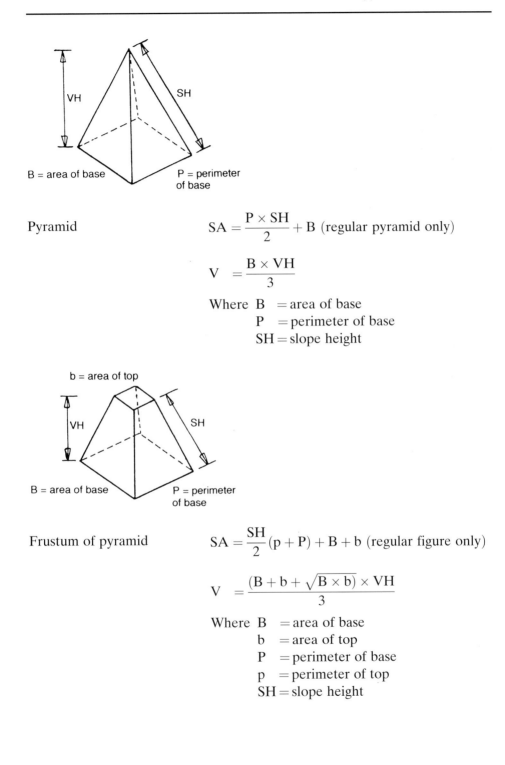

Pyramid

$$SA = \frac{P \times SH}{2} + B \ \text{(regular pyramid only)}$$

$$V = \frac{B \times VH}{3}$$

Where B = area of base
P = perimeter of base
SH = slope height

Frustum of pyramid

$$SA = \frac{SH}{2}(p + P) + B + b \ \text{(regular figure only)}$$

$$V = \frac{(B + b + \sqrt{B \times b}) \times VH}{3}$$

Where B = area of base
b = area of top
P = perimeter of base
p = perimeter of top
SH = slope height

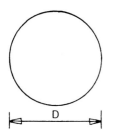

Sphere

$$SA = \pi \times D \times D$$
$$V = 0.5236 \times D \times D \times D$$

$$\left(\text{Note } \frac{\pi}{6} = 0.5236\right)$$

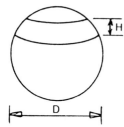

Zone of sphere

$$SA = \pi \times D \times H \text{ (excluding base \& top)}$$

$$V = \frac{\pi \times H}{6} \times (3 \times R \times R + 3 \times r \times r + H \times H)$$

Where R = radius at base
r = radius at top

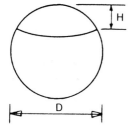

Segment of sphere

$$SA = \pi \times D \times H \text{ (excluding base)}$$

$$V = \frac{\pi \times H}{6} \times (3 \times R \times R + H \times H)$$

Where R = radius of base

Irregular areas

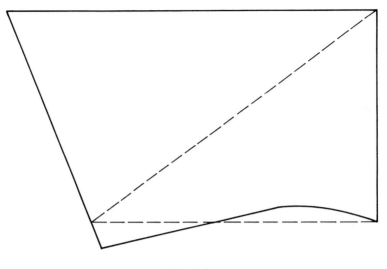

Fig. A.1

Any irregular-shaped area to be measured is usually best divided up into triangles, the triangles being measured individually and then added to give the area of the whole. If one of the sides, as for instance in the case of paving, is irregular or curved, the area can still be divided into triangles by the use of a *compensating*, or *give and take*, line, i.e. a line is drawn along the irregular or curved boundary in such a position that, so far as can be judged, the area of paving excluded by this line is equal to the area included beyond the boundary. In Fig. A.1 the area of paving to be measured is enclosed by firm lines, the method of forming two triangles (the sum of the areas of which equals the whole area) being shown by broken lines.

For a more accurate calculation of the irregular area, particularly if evenly spaced offsets are available dividing the area into an even number of strips, Simpson's rule may be applied. The intermediate offsets should be numbered as it is necessary to distinguish the odd numbers from the even. The formula is given on page 324.

Where two sides of a four-sided figure are parallel to form a trapezoid, it is not necessary to divide the figure into triangles, as the area equals the length of the perpendicular between the parallel sides multiplied by the mean length between the diverging

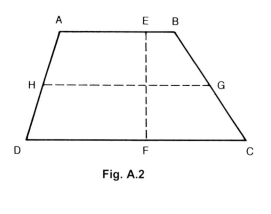

Fig. A.2

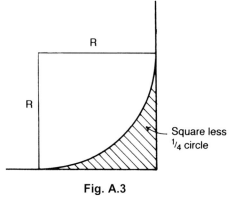

Fig. A.3

sides. In Fig. A.2 the area is $EF \times GH$, GH being drawn halfway between AB and CD and being equal to

$$\frac{AB + CD}{2}$$ i.e. the average of AB and CD

Another irregular figure that often puzzles the beginner is the additional area to be measured where two roads meet with the corners rounded off to a quadrant or bellmouth. This is most easily calculated as a square on the radius with a quarter circle deducted. For example, consider Fig. A.3.

$$\text{Additional area} = R^2 - \tfrac{1}{4} \pi R^2$$

$$= R^2 - \frac{11}{14} R^2$$

$$= \frac{3}{14} R^2$$

Excavation to banks

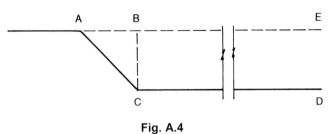

Fig. A.4

The volume of excavation necessary on a level site to leave a regular sloping bank is the sectional area of the part displaced (triangle ABC in Fig. A.4) multiplied by the length; the volume of the remaining excavation equals the sectional area of the rectangle BCDE also multiplied by the length. It may be, however, that the natural ground level is falling in the length of the bank, as shown in Fig. A.5. The volume of earth to be displaced should then theoretically be calculated by the prismoidal formula. As the final volume is required in cubic metres the calculation is carried out in metres:

$$V = \frac{L(A + a + 4m)}{6}$$

Where:

V = Volume
L = Length
A = Sectional area at one end
a = Sectional area at the other end
m = Sectional area at the centre.

Note that in Fig. A.5 the area m is not the average of areas A and a, but should be calculated from the average dimensions.

If AB and BC in Fig. A.4 are both 2.00 m at the higher end and 1.00 m at the lower end, they would, assuming a regular slope, be 1.50 m at the centre. The volume of the prismoid in Fig. A.5 in cubic metres would therefore be:

$$\frac{15 \left(\frac{2 \times 2}{2} + \frac{1 \times 1}{2} + 4 \times \frac{1.5 \times 1.5}{2} \right)}{6}$$

$$= \frac{15 \left(2 + 0.50 + 4.50 \right)}{6} = \frac{105}{6}$$

$$= 17.50 \, \text{m}^3$$

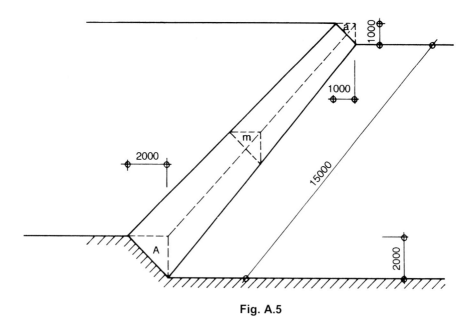

Fig. A.5

In practice, however, it will often be found that so precise a
calculation is not made in such cases. The surface of ground,
whether level or sloping, is not like a billiard cloth, and the natural
irregularities prevent the calculation from being exact. Moreover,
in dealing with normal construction sites the excavation for banks
is a comparatively small proportion of the whole excavation
involved (unlike, say, the case of a railway cutting); any departure
from strict mathematical accuracy due to the use of less precise
methods would involve a comparatively small error. In the
example given above the volume might in practice be taken as the
length multiplied by the sectional area at the centre, i.e.

$$15 \times \frac{1.50 \times 1.50}{2} = 16.88 \, \text{m}^3$$

Although an error of $0.62 \, \text{m}^3$ may be thought high, it must be
remembered that it is being assumed that the excavation to the
bank is only a small proportion of the whole.

If the natural ground level falls to such an extent that it reaches
the reduced level, and the bank shown in Fig. A.5 therefore dies
out to nothing (as in Fig. A.6), the formula will be found to simplify
itself. a becomes zero. m becomes by simple geometry $\frac{A}{4}$, because
the triangles at A and m are right-angled triangles, that at m having

two sides enclosing the right angle each half the length of the corresponding sides of the triangle at A.

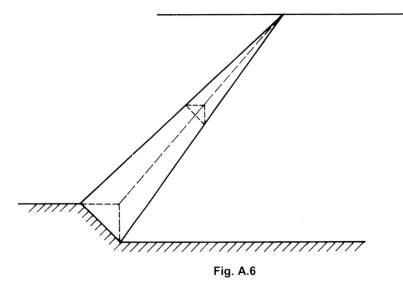

Fig. A.6

$$V = L\frac{(A + 0 + 4 \times \frac{A}{4})}{6} = \frac{L \times A}{3}$$

which is the formula for the volume of a pyramid.

If the volume of earth to be excavated forms an even number of prismoids of equal length, then Simpson's rule may be applied taking the area at the offsets rather than the length as in the case of the irregular area in Fig. A.5.

Excavation to sloping sites

The theoretical principle for measuring the volume of excavation in cutting for a sloping bank may be extended to a sloping site (Fig. A.7). If the figures given in the diagram are natural levels and it is required to excavate to a general level of +1 (the end boundaries of the area being parallel), the volume of excavation may be calculated by the prismoidal formula already given:

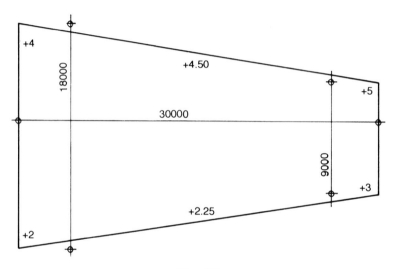

Fig. A.7

$$V = \frac{L(A + a + 4m)}{6}$$

When:

$L = 30$

$$A = 18 \times \frac{3+1}{2} = 36$$

$$a = 9 \times \frac{4+2}{2} = 27$$

$$m = 13.50 \times \frac{3\frac{1}{2} + 1\frac{1}{4}}{2} = 13.50 \times \frac{19}{4} \times \frac{1}{2} = \frac{256.50}{8}$$

then:

$$V = \frac{30(36+27) + \frac{256.50 \times 4}{8}}{6} = \frac{30\ (36+27+128.25)}{6}$$

$$= 5 \times 191.25 = 956.25\,\text{m}^3$$

If only a simple average of the four corners were taken, the result would be:

$$
\begin{array}{r}
4 \\
5 \\
3 \\
2 \\
\hline
\div\ 4)\overline{14} \\
\end{array}
$$

Average ground level	3.50
Reduced level	1.00
Average depth	2.50

$$30 \times 13.50 \times 2.50 = 1012.5\,\text{m}^3$$

This shows an error that is greater because the intermediate level on the lower boundary is not an average of the two end levels.

An average of the eight levels (giving the centre levels double value) would be:

$$
\begin{array}{rr}
& 4 \\
2/4.50 = & 9 \\
& 5 \\
& 3 \\
2/2.25 = & 4.50 \\
& 2 \\
\hline
\div\ 8)\overline{27.50} \\
& 3.44 \\
& -1.00 \\
\hline
& 2.44 \\
\end{array}
$$

or a volume of $30 \times 13.50 \times 2.44 = 988.2\,\text{m}^3$, from which it will be seen that there is a definite error. This method would be satisfactory, however, if the area of the ground were a rectangle.

The error would vary with the regularity of the slope, and where the slope is fairly regular it may often be sufficiently accurate to take an average depth over the whole area. For an ordinary site, calculation would probably be made in this way in practice.

If the formula is to be applied where end boundaries are not parallel, it will be necessary to draw parallel compensating lines along these two boundaries. It is, of course, assumed that slopes between the levels given are regular. If a line of intermediate levels were given between the upper and lower lines in Fig. A.7, it would be necessary to treat the two portions separately.

Grids of levels

On larger sites it is customary to take a grid of levels over the whole area, or at least over the area of the proposed construction, at regular intervals forming squares, plus additional levels at any significant points such as existing manhole covers or along embankments. When calculating the average level of an area covered by a grid, it is necessary to find the average level of each square of the grid by totalling the levels at the four corners and dividing by four. The average levels thus found are added together, the total is divided by the number of squares and the result gives the average level of the area. To calculate the amount of excavation the total area is multiplied by the difference between the average level of the area and the formation level required. If the excavation of top soil has been measured as a separate item then the depth of this should be deducted from the average level.

A quicker method is to total the levels in each of the categories indicated below and multiply each total by the appropriate *weighting* shown.

(1) Levels at the external corners on the
boundary of the area   Multiply by 1
(2) Levels at the boundaries of the area
(other than those in (1) or (3))   Multiply by 2
(3) Levels at any internal corners on the
boundary of the area   Multiply by 3
(4) Intermediate levels within the area   Multiply 4

The results obtained are totalled and divided by a number equal to four times the number of squares (or the total number of weightings). The result is the average level of the area and will produce the same result as the first method.

Example of weighted average excavation

Assume a grid of 4 squares with each point 5 m apart, as shown in Fig. A.8. The existing levels are included in the table and the formation that this needs to be excavated down to is at 62.000.

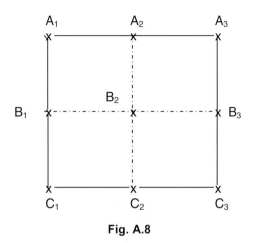

Fig. A.8

Existing ground levels

	× 1		× 2		× 4
A_1	62.250	A_2	62.750	B_2	63.500
A_3	63.100	B_1	62.900		
C_1	63.500	B_3	63.800		
C_3	64.100	C_2	63.800		
1 ×	252.950	2 ×	253.250		

506.500	254.000
	506.500
	252.950
	16)1 013.450

weighted average ground level	=	63.341
less formation level	=	62.000
Depth of reduced level excavation		1.341

This would now be transferred to the dimension paper as follows:

4/	5.00	Excavate to reduce levels,
	5.00	maximum depth ≤ 2 m
	<u>1.34</u>	

Sometimes a site may be partly excavated and partly filled and it will be necessary to plot a cut and fill contour using interpolation, as shown below, to ascertain the location. Contours may also have to be plotted to represent the division between depth bands of excavation or fill as required by the SMM. Grid squares that are cut by the contour line to form triangles or trapezia should be dealt with separately; the average level of each is found by adding the levels at the corners and dividing by the number of corners. The area of each irregular figure is multiplied by the difference between its average level and the formation level to give the volume of excavation or fill to add to the main quantity.

When measuring excavation work, one should always look for sudden changes in levels which may indicate items such as embankments or craters. These areas should be dealt with separately and not averaged within the main area as above.

Interpolation of levels

When measuring excavation work, it is sometimes necessary to ascertain the ground level at a point between two levels given on the drawing or to locate a point at which a certain level occurs. This can be achieved by interpolation as shown in the following example:

Given two levels at points A and B 50 m apart, find an intermediate level at point C 20 m from A.

Level at A = 3.60
Level at B = <u>2.90</u>
Difference   0.70
Therefore the ground falls 0.70 over 50 m
To find the fall y over 20 m

$$\frac{0.70}{50} = \frac{y}{20}$$

$20 \times 0.70 = 50y$

$y = 0.28$

The level at $C = 3.60 - 0.28 = \underline{3.32}$

If the intermediate level is known, such as a formation level, then the distance from A could be found in a similar manner, the unknown being the distance rather than the level.

Index